Switched Inductors:

Building Blocks

By

Gabriel Alfonso Rincón-Mora

School of Electrical and Computer Engineering
Georgia Institute of Technology

Rincon-Mora.gatech.edu

Copyright © 2021 by G.A. Rincón-Mora

Copyright © 2021 by Gabriel A. Rincón-Mora. All rights reserved.

No part of this publication may be reproduced, stored in a retrieval system or transmitted in any form or by any means, electronic, mechanical, photocopying, recording, scanning or otherwise, except as permitted under Section 107 or 108 of the 1976 United States Copyright Act, without the prior written permission of the author.

Switched Inductors: Building Blocks

Contents

	Page
List of Figures	v
List of Abbreviations	vii
1. Current Sensors	**1**
1.1. Series Resistance	1
A. Sense Resistor	1
B. MOS Resistance	2
C. Inductor Resistance	3
1.2. Sense Transistor	7
A. Sense FET	7
B. Looped Sense FET	9
1.3. Design Notes	10
2. Voltage Sensors	**11**
2.1. Voltage Divider	11
2.2. Phase-Saving Voltage Divider	13
2.3. Voltage-Dividing Error Amplifier	14
3. Digital Blocks	**16**
3.1. Push-Pull Logic	16
A. Inverter	16
B. NOR Gate	18
C. NAND Gate	19
D. Design Notes	19
3.2. SR Flip Flops	20
3.3. Gate Driver	21
A. Minimum Delay	22
B. Gate-Charge Power	24
C. Shoot-Through Power	25
D. Design Notes	27
3.4. Dead-Time Logic	28
4. Comparator Blocks	**30**
4.1. Comparators	30
4.2. Hysteretic Comparators	30
4.3. Summing Comparators	31
5. Timing Blocks	**32**
5.1. Clocked Sawtooth Generator	32
5.2. Sawtooth Oscillator	34
5.3. One-Shot Oscillator	36

6. Switch Blocks — 37
- 6.1. Class-A Inverters — 37
- 6.2. Supply-Sensing Comparators — 39
  - A. Low Side — 39
  - B. High Side — 39
  - C. Offset — 40
  - D. Design Notes — 41
- 6.3. Zero-Current Detectors — 42
- 6.4. Ring Suppressor — 42
- 6.5. Switched Diodes — 44
  - A. Low Side — 44
  - B. High Side — 45
- 6.6. Starter — 45
  - A. Shutdown — 45
  - B. Startup — 47

7. Summary — 48

List of Figures

		Page
Figure 1.	Sense resistor.	1
Figure 2.	MOS resistance.	2
Figure 3.	Low-pass-filtered inductor resistance.	3
Figure 4.	Bypass-filtered inductor resistance.	4
Figure 5.	Low- and high-side sense transistors.	7
Figure 6.	Complementary sense transistors.	9
Figure 7.	Looped low- and high-side sense transistors.	9
Figure 8.	Looped complementary sense transistors.	10
Figure 9.	Voltage divider.	12
Figure 10.	Phase-saving divider.	13
Figure 11.	Translating error amplifier.	15
Figure 12.	Push–pull inverter.	16
Figure 13.	Push–pull NOR gate.	18
Figure 14.	Push–pull NAND gate.	19
Figure 15.	SR flip flop.	20
Figure 16.	Analog set/reset-dominant flip flops.	20
Figure 17.	Digital set/reset-dominant flip flops.	21
Figure 18.	Gate driver.	21
Figure 19.	Optimal gate-driver setting for minimum delay.	23
Figure 20.	Inverter's shoot-through response.	25
Figure 21.	Dead-time response.	29
Figure 22.	Active-high and -low dead-time circuits.	29
Figure 23.	Typical comparator.	30
Figure 24.	Flip-flopped hysteretic comparator.	31
Figure 25.	Summing comparator.	32
Figure 26.	Summing hysteretic comparator.	32
Figure 27.	Clocked sawtooth generator.	33
Figure 28.	Sawtooth oscillator.	34
Figure 29.	One-shot oscillator.	36
Figure 30.	Class-A inverter.	38
Figure 31.	Positive and negative low-side comparators.	39
Figure 32.	Positive and negative high-side comparators.	40
Figure 33.	Low- and high-side zero-current detectors.	42
Figure 34.	Ring suppressor.	43

Figure 35.	Switched low-side diode.	45
Figure 36.	Switched high-side diode.	45
Figure 37.	Shutdown buck–boost.	46
Figure 38.	Shutdown buck.	46
Figure 39.	Shutdown boost.	46

List of Abbreviations

ADC ≡ Analog–digital converter
DCM ≡ Discontinuous-conduction mode
EMI ≡ Electromagnetic interference
ICMR ≡ Input common-mode range
LED ≡ Light-Emitting Diode
OCP ≡ Over-current protection
PM ≡ Phase margin
SL ≡ Switched Inductor
SR ≡ Set–Reset
ZCD ≡ Zero-current detector

A_0 ≡ Low-frequency gain
A_β ≡ Feedback translation
A_E ≡ Error amplifier
A_F ≡ Forward translation
A_{IE} ≡ Current-error amplifier
A_{LG} ≡ Loop gain
A_M ≡ Mirror gain
A_S ≡ Stabilizing filter response

β_{IFB} ≡ Current-feedback translation
β_{FB} ≡ Feedback translation

C_{CH} ≡ Channel capacitance
C_D ≡ Drain capacitance
C_{EI} ≡ A_E's input capacitance
C_G ≡ Gate capacitance
C_{GI} ≡ Input gate capacitance
C_{GO} ≡ Output gate capacitance
C_O ≡ Output capacitance
$C_{OX}"$ ≡ Oxide capacitance (per unit area)
CP_{IE} ≡ Current-error comparator

D_{DG} ≡ Ground drain diode
D_{DO} ≡ Output drain diode
d_E' ≡ Duty-cycle command
D_H ≡ High-side diode
D_L ≡ Low-side diode

f_{0dB} ≡ Unity-gain frequency
f_{LC} ≡ LC frequency
f_o ≡ Fan out
F_O ≡ Total fan out

$f_{RC} \equiv$ RC frequency
$f_{RL} \equiv$ RL frequency

$I_B \equiv$ Bias current
$i_C \equiv$ Collector current
$i_E \equiv$ Error current
$i_{FB} \equiv$ Feedback current
$i_L \equiv$ Inductor current
$i_{L(D)} \equiv$ Inductor drain current
$i_O \equiv$ Output current
$i_P \equiv$ Power current
$i_S \equiv$ Sense current
$I_S \equiv$ Saturation current
$i_{ST} \equiv$ Shoot-through current

$k_{SL} \equiv$ Self-loading coefficient

$L \equiv$ Length
$L_{OL} \equiv$ Overlap length
$L_X \equiv$ Switched inductor
$\lambda \equiv$ Channel-length modulation parameter

$M_H \equiv$ High-side switch
$M_L \equiv$ Low-side switch
$M_P \equiv$ Power transistor
$M_S \equiv$ Sense transistor

$p_A \equiv$ Amplifier pole
$P_{DRV} \equiv$ Gate-driver power
$P_{IV} \equiv i_{DS}-v_{DS}$ overlap power
$P_G \equiv$ Gate-charge power
$P_Q \equiv$ Quiescent power
$P_{ST} \equiv$ Shoot-through power

$Q \equiv$ Current state
$\overline{Q} \equiv$ Complementary state
$Q^{-1} \equiv$ Previous state
$q_G \equiv$ Gate charge

$R_D \equiv$ Drain resistance
$R_{DS} \equiv$ Drain–source resistance
$R_L \equiv$ Inductor resistance
$R_{LD} \equiv$ Load resistance
$R_S \equiv$ Series resistor

$\Sigma v_{ID} \equiv$ Differential sum

t_{CLK} ≡ Clock period
t_D ≡ Drain time
t_{DT} ≡ Dead time
t_I ≡ Inverter-chain delay
t_E ≡ Energize time
t_{LC} ≡ LC period
t_{OSC} ≡ Oscillation period
t_P ≡ Propagation delay
t_{PW} ≡ Pulse width
t_R ≡ Response/reset time
t_S ≡ Suppression time
t_{ST} ≡ Shoot-through time
t_{SW} ≡ Switching period
τ_{LC} ≡ LC time constant
τ_{RC} ≡ RC time constant

v_B ≡ Base voltage
v_{BE} ≡ Base–emitter voltage
V_{BG} ≡ Bandgap voltage
v_{DD} ≡ Positive power supply
$v_{DS(SAT)}$ ≡ Saturation voltage
v_E ≡ Error voltage
v_{FB} ≡ Feedback voltage
v_G ≡ Gate voltage
v_H ≡ High input
v_I ≡ Input
v_{ID} ≡ Differential input voltage
v_{IFB} ≡ Current-feedback voltage
v_{IN} ≡ Input voltage
v_L ≡ Inductor voltage/low input
v_n ≡ Noise voltage
v_N ≡ Negative input
v_O ≡ Output/output voltage
$V_{OS(S)}$ ≡ Systemic offset voltage
v_P ≡ Positive input
v_R ≡ Reference voltage
v_S ≡ Sawtooth voltage
v_{SS} ≡ Negative power supply
v_{SWI} ≡ Input switching voltage
v_{SWO} ≡ Output switching voltage
v_T ≡ Trip point
V_t ≡ Thermal voltage
V_{T0} ≡ Zero-bias threshold voltage
$v_{T(HI)}$ ≡ Rising threshold
$v_{T(LO)}$ ≡ Falling threshold

v_{VOS} ≡ Voltage-loop offset

W ≡ Width

Z_{LD} ≡ Load impedance

Switched Inductors: Building Blocks

Switched-inductor (SL) power supplies draw and deliver power. They are *regulators* when *feedback controllers* switch them so their outputs follow a target. *Voltage regulators* regulate voltage, *battery chargers* and *light-emitting diode* (LED) *drivers* regulate current, *battery-charging voltage regulators* regulate both, and *energy harvesters* regulate power.

Good power supplies are *efficient* and *accurate*. Efficient supplies lose a small fraction of the power they draw. And accurate supplies deliver the rest with a voltage or current that is very close to a target. Supplying power with a preset voltage or current like this is a form of *power conditioning*.

Power conditioning requires several actions. Some of these are analog in nature, others are digital, and those in between are mixed-mode. The building blocks that power supplies use reflect this diversity.

1. Current Sensors

1.1. Series Resistance

A. Sense Resistor

Inserting a *series resistor* R_S into the conduction path is one way of sensing current. R_S in Fig. 1 can be in series with the *switched inductor* L_X, the *output* v_O, or the *load* Z_{LD}. In all cases, the *current-feedback translation* β_{IFB} that senses v_O's *output current* i_O or L_X's *inductor current* i_L is R_S:

$$\beta_{IFB} \equiv \frac{V_{IFB}}{i_{L/O}} = \frac{i_{L/O} R_S}{i_{L/O}} = R_S. \tag{1}$$

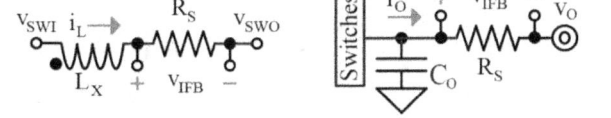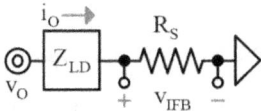

Fig. 1. Sense resistor.

The challenge with adding R_S is the ohmic power R_S dissipates. This is why R_S is usually about or below 1 Ω. The *current-feedback voltage* v_{IFB} that R_S generates is therefore very low. This is unfortunate because the *current-error amplifier* or *comparators* A_{IE} or CP_{IE} requires additional *quiescent power* P_Q when discerning millivolt signals from the noise the switching network generates.

B. MOS Resistance

Sensing resistances already in the conduction path saves the ohmic power that adding R_S burns. But the only resistances accessible are the switches', and they conduct only when they close. So they can only sense part of i_L or i_O. And since these MOS triode *drain–source resistances* R_{DS} are usually very low, A_{IE} or CP_{IE} requires more P_Q when distinguishing the v_{IFB} they generate in Fig. 2 from the surrounding switching noise:

$$\beta_{IFB} \equiv \frac{v_{IFB}}{i_{L/O}}\bigg|_{t_{ON}} = \frac{i_{L/O} R_{DS}}{i_{L/O}}\bigg|_{t_{ON}} = R_{DS}\big|_{t_{ON}}. \quad (2)$$

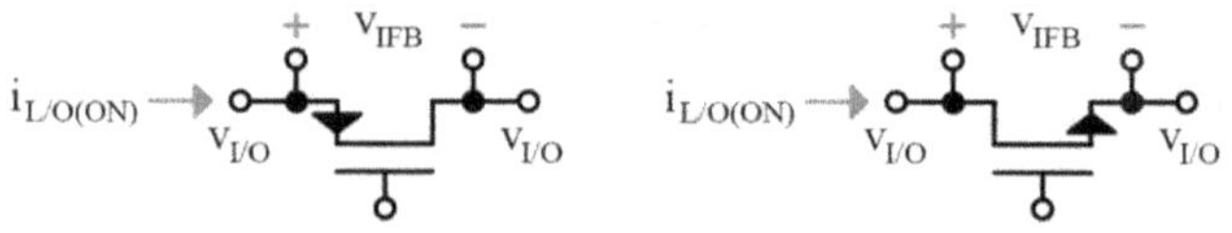

Fig. 2. MOS resistance.

Reconstructing i_L across the *switching period* t_{SW} is possible by sensing switches that conduct across complementary periods. An energize switch can translate i_L across L_X's *energize time* t_E and a drain switch across L_X's *drain time* t_D:

$$\beta_{IFB} \equiv \frac{v_{IFB}}{i_L} = \frac{v_{IFB}\big|_{t_E} + v_{IFB}\big|_{t_D}}{i_L} = R_{DS(E)}\big|_{t_E} + R_{DS(D)}\big|_{t_D}. \quad (3)$$

Superimposing their v_{IFB}'s, however, usually requires more processing, and as a result, more silicon area and power. Luckily, sensing all components of i_L is not always necessary.

C. Inductor Resistance

L_X's *inductor resistance* R_L is also in the conduction path. Except, R_L is *in* L_X, so R_L is not physically accessible. But since L_X shorts at low frequency, low-pass filtering the *inductor voltage* v_L removes the effects of L_X on current and voltage.

R_F's and C_F's in Fig. 3, for example, low-pass filter the *input and output switching voltages* v_{SWI} and v_{SWO} that drop v_L. The voltage between C_F's is therefore the difference between v_{SWI}'s and v_{SWO}'s averages. Since L_X shorts and C_F's open at low frequency, this v_{IFB} is R_L's dc voltage $v_{L(DC)}$. So v_{IFB} carries the voltage i_L's average drops across R_L's dc component:

$$\beta_{IFB} \equiv \frac{v_{IFB}}{i_{L(AVG)}} = \frac{v_{SWI(AVG)} - v_{SWO(AVG)}}{i_{L(AVG)}} \approx \frac{v_{L(DC)}}{i_{L(AVG)}} = \frac{i_{L(AVG)} R_{L(DC)}}{i_{L(AVG)}} = R_{L(DC)}. \quad (4)$$

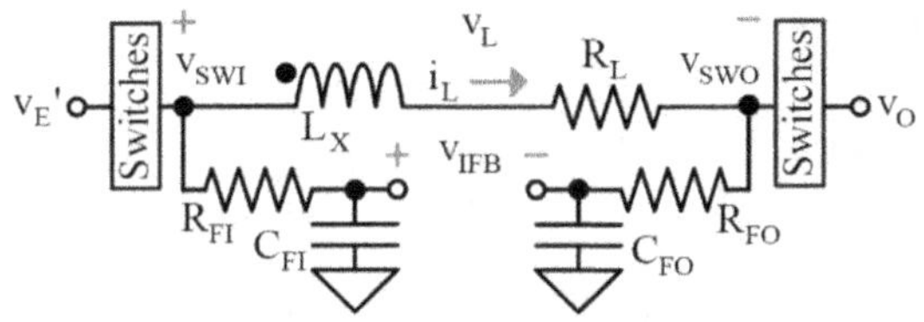

Fig. 3. Low-pass-filtered inductor resistance.

R_L is normally low, so A_{IE} or CP_{IE} requires more P_Q when processing the low v_{IFB} that results. But for this, R_F's and C_F's should first suppress the switching noise v_{SWI} and v_{SWO} produce to a fraction of the voltage R_L drops. This is important because v_{SWI} swings across the *input voltage* v_{IN} and v_{SWO} across the *output voltage* v_O.

To suppress these wide swings, the impedance of C_F should be much lower than R_F. When this happens, Δv_{SWI} slews C_{FI} with v_{IN}/R_{FI} across t_E and Δv_{SWO} slews C_{FO} with v_O/R_{FO} across t_D. The *noise* v_n this generates should be a small fraction of the $v_{L(DC)}$ that $i_{L(AVG)}$ drops across $R_{L(DC)}$:

$$v_n \approx \frac{i_{FI/O} t_{E/D}}{C_{FI/O}}$$

$$\approx \left(\frac{\Delta v_{SWI/O}}{R_{FI/O}}\right)\left(\frac{d_{E/D}t_{SW}}{C_{FI/O}}\right) \approx \left(\frac{v_{IN/O}t_{SW}}{R_{FI/O}C_{FI/O}}\right)\left(\frac{V_{D/E}}{V_E + V_D}\right) \ll v_{L(DC)}. \quad (5)$$

Boosts do not need R_{FI} or C_{FI} because L_X connects directly to v_{IN} and v_{IN} is steady. *Bucks* do not need R_{FI} or C_{FI} for similar reasons, because L_X connects directly to v_O and the *output capacitance* C_O averages v_O. Either way, like their *buck–boost* relative, β_{IFB} only senses i_L's average $i_{L(AVG)}$.

Example 1: Determine C_F for a buck–boost so v_n is less than or equal to 10% of $v_{L(DC)}$ when v_{IN} is 4 V, v_O is 1.8 V, t_{SW} is 1 μs, R_F is 500 kΩ, $R_{L(DC)}$ is 250 mΩ, and $i_{L(AVG)}$ is 300 mA.

Solution:

$$V_E = v_{IN} = 4 \text{ V and } v_D = v_O = 1.8 \text{ V}$$

$$v_{L(DC)} = i_{L(AVG)}R_{L(DC)} = (300m)(250m) = 75 \text{ mV}$$

$$v_n \approx \left(\frac{v_{IN/O}t_{SW}}{R_F C_F}\right)\left(\frac{V_{D/E}}{V_E + V_D}\right) \leq \frac{v_{L(DC)}}{10}$$

$$\approx \left[\frac{v_{IN/O}(1\mu)}{(500k)C_F}\right]\left(\frac{V_{D/E}}{4+1.8}\right) \leq \frac{75m}{10} \quad \therefore C_F \geq 330 \text{ pF}$$

Note: Two 330-pF occupy substantial silicon area.

The bypass RC filter in Fig. 4 adopts a different approach. In this case, since capacitors complement inductors, C_F and R_F emulate the action of L_X and R_L. In other words, R_F and C_F filter v_L into a voltage v_{IFB} the same way L_X and R_L filter v_L into a current i_L.

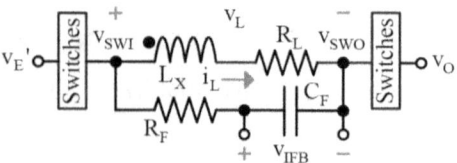

Fig. 4. Bypass-filtered inductor resistance.

i_L is an ohmic translation of v_L across L_X and R_L. v_{IFB} is the voltage a similar ohmic translation of v_L across C_F and R_F drops across C_F:

$$i_L = \frac{v_L}{sL_X + R_L} = \frac{v_L/R_L}{sL_X/R_L + 1} \qquad (6)$$

$$v_{IFB} = \left(\frac{v_L}{1/sC_F + R_F}\right)\left(\frac{1}{sC_F}\right) = \frac{i_L(sL_X + R_L)}{1 + sR_FC_F} = i_L R_L \left(\frac{1 + sL_X/R_L}{1 + sR_FC_F}\right), \qquad (7)$$

where v_L is also the voltage i_L drops across L_X and R_L. So v_{IFB} drops $i_L R_L$ when $R_F C_F$ matches L_X/R_L. In other words, R_L translates i_L to v_{IFB} when R_F and C_F's *RC frequency* f_{RC} matches L_X and R_L's *RL frequency* f_{RL}:

$$f_{RC} = \frac{1}{2\pi R_F C_F} \equiv f_{RL} = \frac{R_L}{2\pi L_X} \qquad (8)$$

$$\beta_{IFB} \equiv \frac{v_{IFB}}{i_L} = \frac{i_L R_L}{i_L} = R_L. \qquad (9)$$

Note this β_{IFB} senses i_L across frequency. But since L_X's *skin effect* crowds i_L towards the edges, R_L in β_{IFB} climbs with frequency. R_L is also low, so A_{IE} or CP_{IE} require additional P_Q to discern the low v_{IFB} that results from the surrounding switching noise.

R_F and C_F can shunt and bypass L_X and R_L. This happens when R_F and C_F's impedance falls below L_X and R_L's. When $R_F C_F$ matches L_X/R_L, R_F and C_F bypass L_X and R_L past f_B when sL_X overcomes R_F:

$$R_F + \frac{1}{sC_F} = R_F + \frac{R_L R_F}{sL_X} = \left(\frac{R_F}{sL_X}\right)(sL_X + R_L)\bigg|_{f_O \geq \frac{R_F}{2\pi L_X} = f_B} \leq sL_X + R_L. \qquad (10)$$

Since L_X and R_F are in μH's and kΩ's, this usually happens at very high frequency. So this unintended effect is largely inconsequential.

Example 2: Determine C_F, $v_{IFB(AVG)}$, and f_B for a buck–boost when R_F is 500 kΩ, L_X is 10 μH, R_L is 250 mΩ, and $i_{L(AVG)}$ is 300 mA.

Solution:

$$R_F C_F = (500k)C_F \equiv \frac{L_X}{R_L} = \frac{10\mu}{250m} \quad \therefore C_F = 80 \text{ pF}$$

$$v_{IFB(AVG)} = i_{L(AVG)} R_L = 75 \text{ mV}$$

$$f_B = \frac{R_F}{2\pi L_X} = \frac{500}{2\pi(10\mu)} = 8.0 \text{ GHz}$$

Matching time constants is not always necessary. After L_X overcomes R_L (past f_{RL}) and C_F shorts with respect to R_F (past f_{RC}), v_L/R_F drops v_{IFB} across $1/sC_F$ with the v_L that i_L drops across sL_X. So $\beta_{IFB(AC)}$ is $L_X/R_F C_F$, which can be less or greater than one:

$$\beta_{IFB(AC)} \equiv \left.\frac{v_{IFB}}{i_L}\right|_{\substack{f_O \gg f_{RL} \\ f_O \gg f_{RC}}} \approx \left(\frac{v_L}{R_F}\right)\left(\frac{Z_C}{i_L}\right) \approx \frac{i_L s L_X}{R_F s C_F i_L} = \frac{L_X}{R_F C_F}. \quad (11)$$

Since L_X shorts and C_F opens at low frequency, $\beta_{IFB(DC)}$ is R_L:

$$\beta_{IFB(DC)} \equiv \left.\frac{v_{IFB}}{i_L}\right|_{\substack{f_O \ll f_{RL} \\ f_O \ll f_{RC}}} \approx \frac{i_L R_L}{i_L} = R_L. \quad (12)$$

$\beta_{IFB(AC)}$ equals R_L when $R_F C_F$ matches L_X/R_L. When $\beta_{IFB(AC)}$ is higher than R_L, β_{IFB} rises over $\beta_{IFB(DC)}$ when sL_X overcomes R_L and reaches $\beta_{IFB(AC)}$ when C_F shorts. So $\beta_{IFB(AC)}$ only applies to frequencies over f_{RL} and f_{RC}.

C_F is $L_X/\beta_{IFB(AC)}R_F$ when setting $\beta_{IFB(AC)}$ to a particular gain. This C_F and R_F bypass L_X and R_L when C_F and R_F's impedance falls below L_X and R_L's. This f_B', however, is higher than f_B when $\beta_{IFB(AC)}$ is greater than R_L:

$$R_F + \frac{1}{sC_F} = R_F + \left.\frac{\beta_{IFB(AC)}R_F}{sL_X}\right|_{f_O \geq f_B\left(\frac{sL_X + \beta_{IFB(AC)}}{sL_X + R_L}\right) = f_B'} \leq sL_X + R_L. \quad (13)$$

Example 3: Determine C_F for Example 2 so $\beta_{IFB(AC)}$ is 10 Ω.

Solution:

$$\beta_{IFB(AC)} \approx \frac{L_X}{R_F C_F} = \frac{10\mu}{(500k)C_F} \equiv 10\ \Omega \quad \therefore\ C_F = 2\ pF$$

A drawback to this filter is the switching nature of v_{SWI} and v_{SWO}. Designing A_E or CP_E to tolerate this common-mode variation in v_{IFB} is challenging. But since boosts and bucks switch only their input or output, v_{SWI} or v_{SWO} is steady at v_{IN} or v_O. Connecting C_F to the steady voltage anchors v_{IFB}. In buck–boosts, connecting C_F to the least variable v_{SW} eases A_E's or CP_E's *input common-mode range* (ICMR) requirement.

1.2. Sense Transistor

A. Sense FET

Currents of nearby transistors match when their geometries and terminal voltages match. Matched this way, a sense transistor can mirror the current of a power switch across temperature and fabrication runs. The challenge is reproducing the voltage dropped across the switch, which is very low. This is important because transistors that drop millivolts are in triode, where their currents are sensitive to voltage.

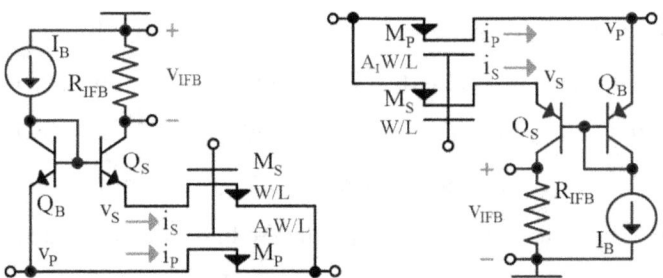

Fig. 5. Low- and high-side sense transistors.

Sense and *power transistors* M_S and M_P in Fig. 5 share gate, source, and body connections and voltages. Q_B carries a steady *bias current* I_B and Q_S carries M_S's *sense current* i_S. Since *base–emitter voltage* v_{BE} is a logarithmic function of *collector current* i_C, Q_S's and Q_B's v_{BE}'s are roughly the same. So Q_S matches M_S's drain voltage v_S to M_P's v_P.

With terminal voltages matched, i_S mirrors M_P's *power current* i_P. To save silicon area and power, M_S is A_I times or 1000× or so smaller than M_P, so i_S is similarly lower than the i_L or i_O that i_P carries. This i_S flows through Q_S and R_S to set a v_{IFB} that scales with i_L or i_O when M_P conducts:

$$\beta_{IFB} \equiv \left.\frac{v_{IFB}}{i_{L/O}}\right|_{t_{ON}} \approx \left.\frac{(i_{L/O}/A_I)R_{IFB}}{i_{L/O}}\right|_{t_{ON}} = \left.\frac{R_{IFB}}{A_I}\right|_{t_{ON}}. \quad (14)$$

Unfortunately, M_S and M_P's matching accuracy suffers when M_S is much smaller than M_P. This is the tradeoff for scaling i_S. R_{IFB} also varies with temperature and fabrication runs. But since M_S tracks M_P and R_{IFB} is usually high, these inaccuracies can be manageable.

Q_S's and Q_B's v_{BE}'s match well when i_S and I_B equal. But since i_S scales with i_L or i_O and I_B is constant, i_S does not always equal I_B. This difference causes a mismatch between Q_S's and Q_B's v_{BE}'s that offsets M_S's and M_P's drain voltages v_S and v_P. This nonlinear *error voltage* v_E distorts β_{IFB}:

$$v_E = v_S - v_D = \Delta v_{BE} = V_t \ln\frac{i_S}{I_B}. \quad (15)$$

Although undesirable, this distortion is not always prohibitive.

Reconstructing i_L across t_{SW} is possible by sensing complementary switches and mirroring one's i_S into the other's R_{IFB}. M_{LS} in Fig. 6, for example, senses M_L's low-side current and M_{M1}–M_{M2} mirrors i_{LS} into M_H's high-side R_{IFB}. This way, R_{IFB}'s v_{IFB} scales with i_{LS} and i_{HS}. Since i_{LS} is zero when i_{HS} is not and *vice versa*, v_{IFB} scales with the i_L that i_{LP} and i_{HP} feed. M_{M1}–M_{M2}'s *mirror gain* A_M ensures i_{LS}'s and i_{HS}'s scaling factors A_M/A_{LI} and $1/A_{HI}$ match:

$$\beta_{IFB} \equiv \frac{v_{IFB}}{i_L} \approx \frac{(i_{LS}A_M + i_{HS})R_{IFB}}{i_L} \approx \left(\frac{i_{L(L)}A_M}{A_{LI}} + \frac{i_{H(L)}}{A_{HI}}\right)\left(\frac{R_{IFB}}{i_L}\right) = \frac{R_{IFB}}{A_I}. \quad (16)$$

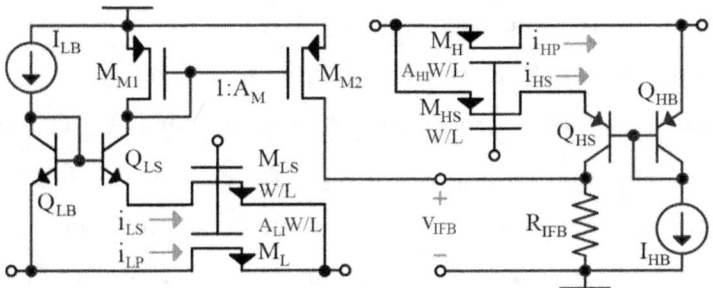

Fig. 6. Complementary sense transistors.

B. Looped Sense FET

The feedback loops in Fig. 7 eliminate β_{IFB}'s distortion. This is because, with equal I_B's, Q_S's and Q_B's i_C's and v_{BE}'s match. This way, v_S follows v_P.

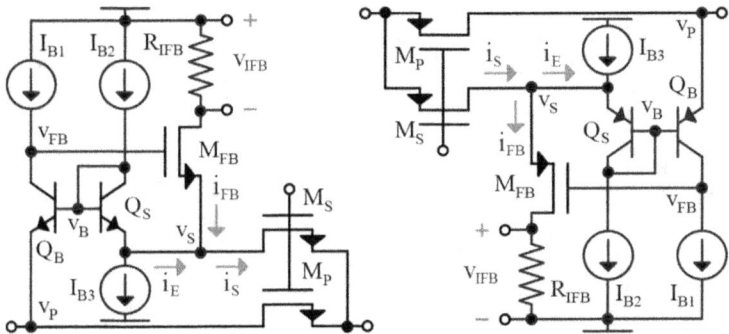

Fig. 7. Looped low- and high-side sense transistors.

Q_S, Q_B, and M_{FB} close a feedback loop that compares i_S with *feedback current* i_{FB}. Resistance at v_S converts the *error current* i_E into a v_S that Q_S level-shifts to *base voltage* v_B and Q_B and M_{FB} amplify and translate into i_{FB}. With enough *loop gain* A_{LG}, i_E is low, i_{FB} nears i_S, and i_{FB} flows through M_{FB} into R_{IFB} to set v_{IFB}. Even when offset by a small mismatch between I_{B2} and I_{B3}, the feedback loop scales i_{FB} with i_S without v_S–v_P variations.

The loop is usually inherently stable because the resistance at v_{FB} is much higher than v_S's and v_B's. So with comparable capacitances, the pole at v_{FB} is much lower than at v_S and at v_B. A_{LG} therefore drops past p_{FB} and reaches the *unity-gain frequency* f_{0dB} near or below p_S and p_B.

The challenge is the bandwidth that f_{0dB} sets. When M_P switches, v_S requires time to match v_P. This *response time* t_R, which f_{0dB} sets, should be a fraction of the t_E or t_D across which M_P conducts. Small Q_S, Q_B, and M_{FB} geometries (for low capacitance) and high I_B's help shorten t_R (extend f_{0dB}).

Reconstructing i_L across t_{SW} is possible by sensing complementary switches and mirroring one's i_S into the other's R_{IFB}. M_{LS} in Fig. 8, for example, senses M_L's low-side current and M_{M1}–M_{M2} mirrors i_{LS} into M_H's high-side R_{IFB}. This way, R_{IFB}'s v_{IFB} scales with i_{LS} and i_{HS}. Since i_{LS} is zero when i_{HS} is not and *vice versa*, v_{IFB} scales with the i_L that i_{LP} and i_{HP} feed. M_{M1}–M_{M2}'s mirror gain A_M ensures i_{LS}'s and i_{HS}'s scaling factors match.

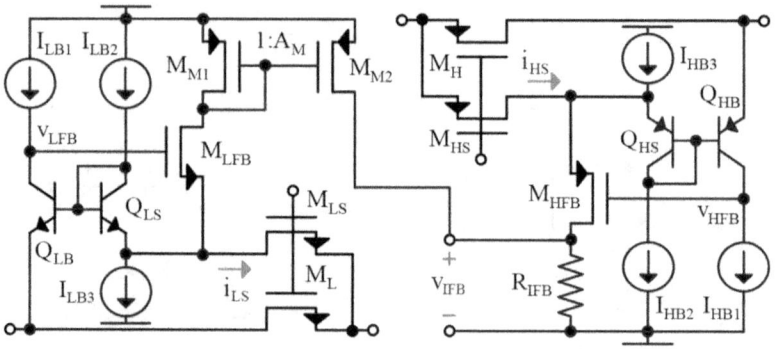

Fig. 8. Looped complementary sense transistors.

1.3. Design Notes

The easiest way of sensing current is by inserting R_S into the conduction path. The problem is, R_S burns power. To keep this power low, R_S should be very low. Discerning the low voltage that R_S drops from the noise the switching network injects is often challenging.

Sensing resistances already in the conduction path save the power that adding R_S burns. Switch resistance is one example. Except, R_{DS}'s only conduct parts of i_L or i_O. And, R_{DS}'s are low and sensitive to gate drive.

R_L is also in the conduction path. R_L, however, is in L_X, switches with v_{SWI} and v_{SWO}, is low, and varies with frequency. Low-pass filters can average v_{SW}'s and extract R_L's dc voltage. A tuned bypass filter can extract R_L's dc and ac voltages, but with a v_{IFB} that swings with one of the v_{SW}'s. This is not a problem for bucks and boosts because they only have one v_{SW}, which the filter can average. This filter can also amplify R_L's ac voltage, but only after L_X overcomes R_L and C_F shorts.

R_S, R_{DS}, and R_L all vary with temperature and fabrication runs. Sense FETs are less sensitive because they track the switches they sense. But since they are much smaller, mismatches offset their outputs. Their translations are also nonlinear. A feedback loop can fix this, but by slowing the response. Although not always, basic sense FETs often offer more favorable tradeoffs than looped FETs and series resistances. Table I summarizes some of these points.

Table I. Current Feedback Translations

	Series Resistances					Sense FETs	
	R_S	R_{DS}	Filtered R_L			Basic	Looped
			Low Pass	Tuned	Untuned		
Input	$i_{O/L}$	$i_{O/L(E/D)}$	$i_{L(DC)}$	i_L	$i_{L(DC/AC)}$	$i_{O/L(E/D)}$	
Gain	R_S	R_{DS}	R_L	R_L	$\dfrac{L_X}{R_F C_F}$	$\dfrac{R_{IFB}}{A_I}$	
Power	$P_R + P_Q$	P_Q	P_Q	P_Q	P_Q	P_Q	
Sensitivity	T_J & Fabrication Runs					Mismatch	
	$R_L \propto f_O$					Nonlinear	Linear

2. Voltage Sensors

2.1. Voltage Divider

v_O is typically over the *bandgap voltage* V_{BG} or sub-V_{BG} that sets the *reference voltage* v_R. So in most cases, the *feedback translation* β_{FB} attenuates v_O to the *feedback voltage* v_{FB} that the *error amp* A_E compares to v_R. This is why v_R is normally 1.2 V or lower and β_{FB} is a fraction.

The *voltage divider* in Fig. 9 is the most common way of translating v_O to v_{FB}. R_{FB1} and R_{FB2} translate v_O into a current i_R that drops v_{FB} across R_{FB2}. β_{FB} is the voltage-divided R_{FB2} fraction of R_{FB1} and R_{FB2}:

$$\beta_{FB} \equiv \frac{v_{FB}}{v_O} = \frac{i_R R_{FB2}}{v_O} = \left(\frac{v_O}{R_{FB1}+R_{FB2}}\right)\left(\frac{R_{FB2}}{v_O}\right) = \frac{R_{FB2}}{R_{FB1}+R_{FB2}}. \quad (17)$$

This ratio is usually accurate because resistors normally match and track well across temperature and fabrication runs.

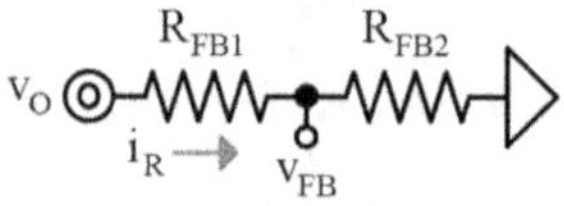

Fig. 9. Voltage divider.

i_R should overwhelm the current noise produces. This i_R and v_O dictate the power P_R that R_{FB1} and R_{FB2} consume. This i_R into R_{FB2} also sets the v_{FB} that reflects v_R and the *voltage-loop offset* v_{VOS}. And with R_{FB1}, R_{FB2} sets v_O's translation to v_{FB}. So when designing the voltage divider, engineers often use v_{FB} and R_{FB2} to set i_R and R_{FB1} to set v_O. This way, R_{FB1} and R_{FB2} account for noise, power, and offset.

Example 4: Determine R_{FB1}, R_{FB2}, and P_R so v_O is 2 V and i_R is 5 µA when v_R is 1.2 V and v_{VOS} is 40 mV.

Solution:

$$P_R = i_R v_O = (5\mu)(2) = 10 \text{ µW}$$

$$i_R = \frac{v_R - v_{VOS}}{R_{FB2}} = \frac{1.2 - 40m}{R_{FB2}} \equiv 5 \text{ µA} \quad \therefore \quad R_{FB2} = 232 \text{ k}\Omega$$

$$\beta_{FB} = \frac{R_{FB2}}{R_{FB1}+R_{FB2}} = \frac{232k}{R_{FB1}+232k}$$

$$= \frac{v_R - v_{VOS}}{v_O} = \frac{1.2 - 40m}{2} = 58\% \quad \therefore \quad R_{FB1} = 168 \text{ k}\Omega$$

2.2. Phase-Saving Voltage Divider

The A_{LG} that controls v_O often includes undesirable poles near or below f_{0dB}. Adding zeros near or below f_{0dB} recovers some of the *phase margin* PM these parasitic poles quench. *Zero–pole pairs* centered about f_{0dB} recover phase at f_{0dB}, where needed, before losing it at higher frequency.

The *phase-saving divider* in Fig. 10 establishes a zero–pole pair z_{FB}–p_{FB} into β_{FB}. C_F should, by design, be much greater than A_E's *input capacitance* C_{EI}, which is parasitic. This way, z_{FB} raises β_{FB} when C_F bypasses R_{FB1}, and since C_O shunts v_O at low frequency, p_{FB} flattens β_{FB} when C_F and C_{EI} shunt R_{FB1} and R_{FB2}:

$$\beta_{FB} \equiv \frac{v_{FB}}{v_O} = \beta_{FB0}\left(\frac{1+s/2\pi z_{FB}}{1+s/2\pi p_{FBX}}\right) = \left(\frac{R_{FB2}}{R_{FB1}+R_{FB2}}\right)\left(\frac{1+s/2\pi z_{FB}}{1+s/2\pi p_{FBX}}\right), \quad (18)$$

$$z_{FB} = \frac{1}{2\pi R_{FB1} C_F}, \quad (19)$$

and $\quad p_{FB} = \dfrac{1}{2\pi(R_{FB1} \| R_{FB2})(C_F + C_{EI})} \approx \dfrac{1}{2\pi(R_{FB1} \| R_{FB2})C_F}. \quad (20)$

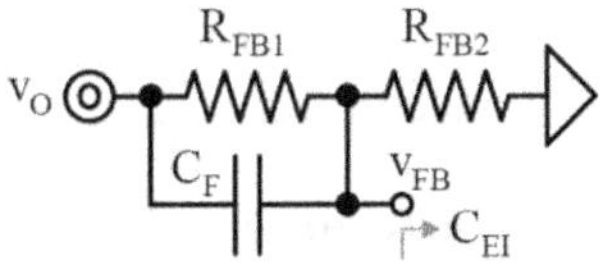

Fig. 10. Phase-saving divider.

C_F is the only open design variable. This is because i_R, v_R, and v_{VOS} determine R_{FB2}, and with R_{FB2}, v_O dictates R_{FB1}. With R_{FB1} and R_{FB2} already set, C_F can center z_{FB} and p_{FB} about f_{0dB} so z_{FB} is as low as p_{FB} is higher than f_{0dB}. How much phase margin C_F saves at f_{0dB} depends on z_{FB} and p_{FB}'s separation, which R_{FB1} and R_{FB2} set:

$$\frac{p_{FB}}{f_{0dB}} = \frac{f_{0dB}}{z_{FB}} \quad (21)$$

$$\Delta PM = \tan^{-1}\frac{f_{0dB}}{z_{FB}} - \tan^{-1}\frac{f_{0dB}}{p_{FB}}. \tag{22}$$

Example 5: Determine C_F and ΔPM for Example 4 when f_{0dB} is 100 kHz.
Solution:

$R_{FB1} = 168\ k\Omega$ and $R_{FB2} = 232\ k\Omega$ from Example 4

$$z_{FB} = \frac{1}{2\pi R_{FB1} C_F} = \frac{1}{2\pi(168k)C_F} = \frac{950n}{C_F}$$

$$p_{FB} = \frac{1}{2\pi(R_{FB1}\|R_{FB2})C_F} = \frac{1}{2\pi(168k\|232k)C_F} = \frac{1.6\mu}{C_F}$$

$$\frac{p_{FB}}{f_{0dB}} = \frac{1.6\mu}{(100k)C_F} \equiv \frac{f_{0dB}}{z_{FB}} = \frac{(100k)C_F}{950n}$$

$$\therefore\ C_F = 12\ pF\ \rightarrow\ z_{FB} = 79\ kHz\ \text{ and }\ p_{FB} = 130\ kHz$$

$$\Delta PM = \tan^{-1}\frac{f_{0dB}}{z_{FB}} - \tan^{-1}\frac{f_{0dB}}{p_{FB}}$$

$$= \tan^{-1}\frac{100k}{79k} - \tan^{-1}\frac{100k}{130k} = 14°$$

2.3. Voltage-Dividing Error Amplifier

When the feedback loop includes A_E, integrating β_{FB} into A_E's stabilizer can save power, components, and area. The *voltage-dividing error amplifier* in Fig. 11, for example, integrates the voltage divider into the feedback network of an *inverting op amp*. R_{FB1} and R_{FB2} translate v_O to v_{FB} and A_V, Z_{F1}, and Z_{F2} with R_{FB1} set A_E's gain and stabilizing response.

Z_{F1} and Z_{F2} are typically capacitors C_{F1} and C_{F2} with, on occasion, series resistors R_{F1} and R_{F2}. This way, Z_{F1} and Z_{F2} open at low frequency. So β_{FB} translates v_O to v_{FB} with R_{FB1} and R_{FB2} and A_E's *low-frequency gain* A_{E0} amplifies v_R and v_{FB}'s error with A_{V0}.

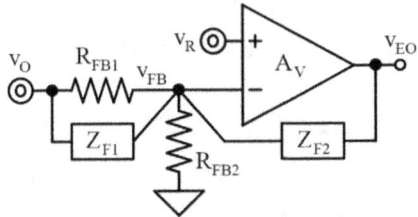

Fig. 11. Voltage-dividing error amplifier.

Z_{F2} closes a feedback loop that alters v_R's and v_O's translations to A_E's output v_{EO}. As the capacitor in Z_{F2} shunts, these gains change in different ways. The stability of the feedback controller depends on v_O's translation to v_{EO}. When v_R is static, v_R's translation to v_{EO} is inconsequential.

The *forward* and *feedback translations* A_F and A_β of the feedback network set the overall gain. A_F excludes the effects of feedback and A_β excludes the effects of limited gain. When combined, the overall gain follows the lowest translation.

In the case of v_O's gain to v_{EO}, A_F voltage-divides with R_{FB1} and R_{FB2} in β_{FB0}, amplifies with A_V's A_{V0}, and filters with Z_{F1}, Z_{F2}, and A_V's *pole* p_A. A_F falls when Z_{F2} shunts R_{FB2}, rises and flattens when Z_{F1} shunts R_{FB1} and shorts with respect to R_{FB1} and R_{FB2}, and falls past p_A:

$$A_F = \left[\frac{R_{FB2} \| Z_{F2}}{(R_{FB1} \| Z_{F1}) + (R_{FB2} \| Z_{F2})}\right]\left(\frac{-A_{V0}}{1+s/2\pi p_A}\right) = -\beta_{FB0}A_{V0}A_X . \quad (23)$$

A_β amplifies with the inverting op-amp gain that Z_{F2} and $R_{FB1} \| Z_{F1}$ set:

$$A_\beta = \frac{-Z_{F2}}{R_{FB1} \| Z_{F1}} . \quad (24)$$

This A_β starts infinitely high and falls with Z_{F2} as C_{F2} shorts.

$\beta_{FB}A_E$ in the feedback controller is v_O's overall gain to v_{EO}. Since this gain follows the lowest translation and A_β's low-frequency gain is infinitely high, $\beta_{FB}A_E$ starts with A_{F0}'s $-\beta_{FB0}A_{V0}$. $\beta_{FB}A_E$ then follows the *stabilizing filter response* A_S that Z_{F1}, Z_{F2}, and p_A set:

$$\beta_{FB}A_E \equiv \frac{V_{EO}}{V_O} = A_F \parallel A_\beta = -\beta_{FB}A_{VO}A_S. \qquad (25)$$

A typical response falls and follows A_β past p_{E1} after A_β falls below A_F. $\beta_{FB}A_E$ can, for example, fall past p_{E1}, flatten when R_{F2} current-limits C_{F2}, climb when C_{F1} bypasses R_{FB1}, and fall when p_A reduces A_F below A_β. Although not necessarily so, engineers normally ensure p_{E1} precedes p_A.

3. Digital Blocks

3.1. Push–Pull Logic

A. Inverter

Inverters invert their *inputs* v_I's. So v_O in Fig. 12 is low or zero when v_I is high or one. And v_O swings high when v_I drops low.

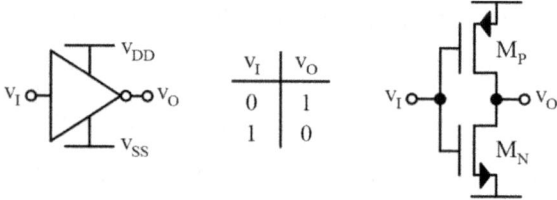

Fig. 12. Push–pull inverter.

The symbol is a triangle. v_I connects to one side and v_O to the small circle on the opposing joint. The flat side blocks static current, the sharp joint drives current, and the small circle inverts polarity.

Digital highs and lows in CMOS implementations usually reach the *positive* and *negative power supplies* v_{DD} and v_{SS}. Since PFETs are active low and NFETs are active high, *push–pull inverters* pull v_O to v_{DD} with a PFET M_P and v_O to v_{SS} with an NFET M_N. Although not always the case, v_{SS} is usually ground.

v_O toggles state when v_I crosses the inverter's *trip point* v_T. v_O holds this state as long as v_I, v_{SS}, and v_{DD} noise keeps v_I above or below v_T. Noise immunity is greatest when v_T is halfway across the supplies.

At this v_T, v_I and v_O are halfway between v_{SS} and v_{DD}. This means, M_P's v_{SG} and v_{SD} and M_N's v_{GS} and v_{DS} are all v_T. With this v_T, M_P and M_N invert their channels when $v_{DD} - v_{SS}$ exceeds M_P's and M_N's *zero-bias threshold voltages* $|V_{TP0}| + V_{TN0}$. They also saturate because, with the same v_T, v_{DS}'s overcome M_P's and M_N's *saturation voltages* $v_{DS(SAT)}$'s, which are a threshold voltage below their v_{GS}'s.

M_P and M_N's *lengths* L should be the minimum L that can sustain v_{DD} and v_{SS}. The *width* W of the stronger device should also be minimum. This way, with the least capacitance possible, v_I and v_O can transition quickly.

At v_T, M_P's and M_N's strengths match. So the width of the weaker device should be large enough to reach the other's strength. In other words, this weaker W should ensure M_P's and M_N's currents match at v_T:

$$\left.\frac{i_P}{i_N}\right|_{v_T}^{v_{GS} > V_{T0}} = \frac{W_P L_N K_P'(v_T - |V_{TP0}|)^2 (1 + v_T \lambda_P)}{W_N L_P K_N'(v_T - V_{TN0})^2 (1 + v_T \lambda_N)} \equiv 1, \qquad (26)$$

where λ's are M_P's and M_N's *channel-length modulation parameters*.

M_P and M_N switch in *sub-threshold* when the voltage across the supplies falls below their thresholds $|V_{TP0}|$ and V_{TN0}. In these cases, M_P and M_N saturate when v_T is three *thermal voltages* V_t's below v_{DD} and $3V_t$ above v_{SS}. With v_T halfway across the supplies, this corresponds to $v_{DD} - v_{SS}$ exceeding $6V_t$. Saturated this way, strengths match when

$$\left.\frac{i_P}{i_N}\right|_{v_T}^{\substack{v_{GS} < V_{T0} \\ v_{DS} > 3V_t}} \approx \left(\frac{W_P L_N I_{SP}}{W_N L_P I_{SN}}\right) \exp\left(V_{TN0} - |V_{TP0}|\right) \equiv 1, \qquad (27)$$

where I_{SP} and I_{SN} are their intrinsic *saturation currents* in sub-threshold.

Example 6: Determine W's when v_{DD} is 4 V, W_{MIN} is 3 μm, L_{MIN}'s match, K_N' is 200 μA/V^2, K_P' is 40 μA/V^2, V_{TN0} is 500 mV, V_{TP0} is − 700 mV, and λ's match.

Solution:

$$v_{DD} - v_{SS} = 4 - 0 > V_{TN0} + |V_{TP0}| = 500m + 700m = 1.2 \text{ V}$$

$$v_T = 0.5 v_{DD} = 0.5(4) = 2 \text{ V}$$

∴ M_P and M_N invert and saturate at v_T

$$\frac{i_P}{i_N} = \frac{W_P L_N K_P'(v_T - |V_{TP0}|)^2}{W_N L_P K_N'(v_T - V_{TN0})^2}$$

$$= \frac{W_P L_N (40\mu)(2 - 700m)^2}{W_N L_P (200\mu)(2 - 500m)^2} = \frac{15 W_P}{100 W_N} \equiv 1$$

∴ $W_N \equiv W_{MIN} = 3 \text{ μm}$ and $W_P = 6.7 W_N = 20 \text{ μm}$

B. NOR Gate

OR gates output high when any of its inputs is high. *NOR* gates invert the action of OR gates. So the *push–pull NOR gate* in Fig. 13 outputs low when any of its inputs is high and high when all inputs are low.

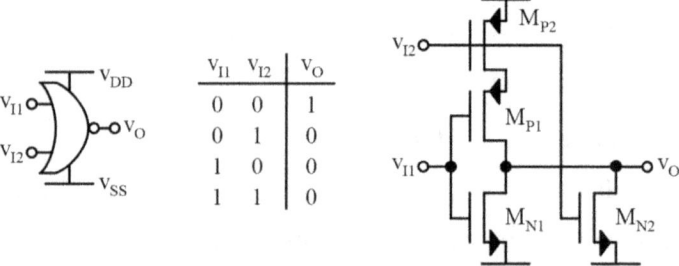

Fig. 13. Push–pull NOR gate.

The NOR gate is basically an inverter with two inputs. Each input requires two transistors. Since v_O is low when any input is high and NFETs are active high, M_{N1} or M_{N2} pulls v_O low when v_{I1} or v_{I2} is high. And M_{P1} and M_{P2} pull v_O high only when v_{I1} and v_{I2} are both low. When split into combinations, parallel transistors OR their inputs and series transistors AND their inputs.

C. NAND Gate

AND gates output a high only when all inputs are high. *NAND* gates invert the action of AND gates. So the *push–pull NAND gate* in Fig. 14 outputs high when any input is low and low when all inputs are high.

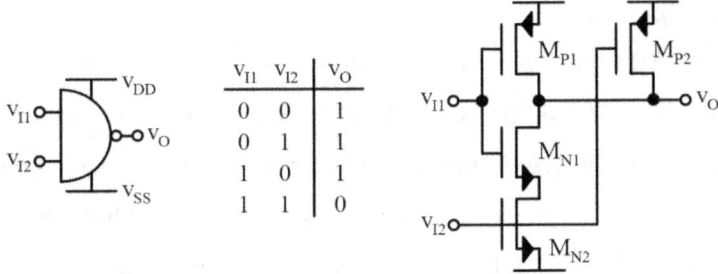

Fig. 14. Push–pull NAND gate.

The NAND gate is basically an inverter with two inputs. Each input requires two transistors. Since v_O is high when any input is low and PFETs are active low, M_{P1} or M_{P2} pull v_O high when v_{I1} or v_{I2} is low. And M_{N1} and M_{N2} pull v_O low only when v_{I1} and v_{I2} are both high. So parallel transistors OR their inputs and series transistors AND their inputs.

D. Design Notes

NFETs and PFETs lose strength when their gate–source voltages are lower than their supplies. M_{P1} in the NOR gate and M_{N1} in the NAND gate are weaker than M_P and M_N in the inverter for this reason, because M_{P2} and M_{N2} reduce their v_{GS}'s. M_{P1}'s v_T is therefore lower than M_{P2}'s and M_{N1}'s v_T is higher than M_{N2}'s.

Although not always necessary, re-centering their v_T's balances their t_P's and noise margins. Since they lose strength to similarly sized devices, doubling the widths of these *source-degenerated transistors* can restore their strength. In other words, v_T's for the inverter, two-input NOR, and two-input NAND circuits are roughly the same when L's and non-degenerated W's match and degenerated W's are twice as wide.

When adding inputs, the concepts used to extend the one-input inverter to the two-input gates still apply. Each additional input requires two FETs: one for the parallel combination and another for the series stack. Three- and four-input variations are not uncommon in systems.

3.2. SR Flip Flops

Set–Reset (SR) *flip flops* "set" or "reset" their *current state* "Q" high or low when their "S" or "R" input in Fig. 15 is high and the other is low. They hold their *previous state* Q^{-1} when both inputs are low. And when both inputs are high, *set-dominant flip flops* output high and *reset-dominant flip flops* output low. Either way, SR implementations normally output Q and Q's *complementary state* $\overline{Q}$.

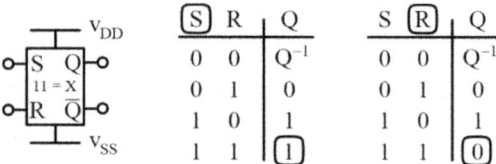

Fig. 15. SR flip flop.

Flip flops perform four functions: set, reset, hold, and dominate. The *analog SR flip flops* in Fig. 16 use set and reset switches to connect Q to v_{DD} and v_{SS} when SR is high–low and low–high. The set resistance R_S in the set-dominant case is much lower than the reset resistance R_R and *vice versa* for the reset-dominant case. This way, Q approaches v_{DD} or v_{SS} when SR is high–high. And capacitor C_H holds Q^{-1} when SR is low–low.

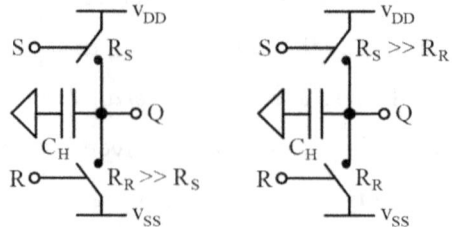

Fig. 16. Analog set/reset-dominant flip flops.

The *digital SR flip flops* in Fig. 17 use set and reset NOR gates to set and reset Q when SR is high–low and low–high. The inverter connects to the set- or reset-dominant NOR gate. This way, Q sets with S or resets with R when SR is high–high. Positive feedback holds Q^{-1} when SR is low–low. Holding Q^{-1} is possible because NOR gates ignore low inputs.

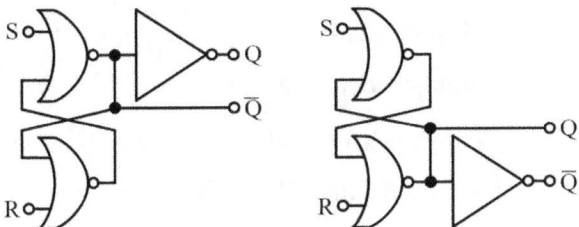

Fig. 17. Digital set/reset-dominant flip flops.

3.3. Gate Driver

Power switches are typically large to limit the power they burn when they conduct the i_L that feeds i_O. So the *output gate capacitance* C_{GO} that *gate drivers* feed is usually very high. C_{GO} is so high that a minimum-size inverter requires too much time to charge and discharge C_{GO}.

The chain of increasingly larger inverters in Fig. 18 can build the current needed to drive C_{GO}. To unload the circuit that feeds the driver, the first stage K_1 should be a minimum-size inverter. This way, the *input gate capacitance* C_{GI} of the driver is the lowest capacitance possible.

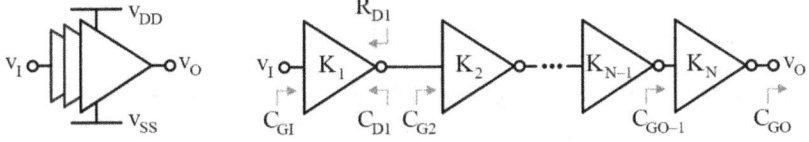

Fig. 18. Gate driver.

Subsequent stages are f_o times greater than their preceding stage. This f_o is the *fan out* because each *gate* load *capacitance* C_G, which scales with W, is f_o times higher than the preceding gate load C_{G-1}. When f_o is consistent across N stages, C_{GO} is $C_{GI} f_o^N$. The *total fan out* F_O is C_{GO}/C_{GI}, which when L's match, is also the ratio of the W_{GO} and W_{GI} that set C_{GO} and C_{GI}. So in short, F_O is f_o^N, or the other way around, f_o is $F_O^{1/N}$:

$$F_O = \frac{C_{GO}}{C_{GI}} = \frac{C_{GI} f_o^N}{C_{GI}} = f_o^N = \left(\frac{C_{GO}}{C_{GO-1}}\right)^N = \frac{W_{GO}}{W_{GI}}. \quad (28)$$

A. Minimum Delay

K_1's *drain resistance* R_{D1} drives K_1's *drain capacitance* C_{D1} and K_2's gate load C_{G2}. K_1's total output capacitance C_{O1} needs 69% of R_{D1} and C_{O1}'s *RC time constant* τ_{RC} to swing v_{O1} halfway across the maximum swing $\Delta v_{O(MAX)}$ that v_{DD} and v_{SS} set. This 69%τ_{RC} is K_1's *propagation delay* t_{P1}:

$$\Delta v_O = \Delta v_{O(MAX)}\left(1 - \exp\frac{-t_\%}{\tau_{RC}}\right) = (v_{DD} - v_{SS})\left(1 - \exp\frac{-t_\%}{\tau_{RC}}\right), \quad (29)$$

$$t_{P1} \equiv t_{50\%\Delta v_{O(MAX)}} = \tau_{RC} \ln(1 - 50\%)^{-1} = 69\%\tau_{RC} = 69\%R_{D1}C_{O1}, \quad (30)$$

and
$$C_{O1} = C_{D1} + C_{G2} = C_{GI}(k_{SL} + f_o) \approx C_{GI}(1 + f_o). \quad (31)$$

k_{SL} relates C_{D1} to C_{GI} because the gate capacitances that set C_{GI} are also present in C_{D1}. Although not necessarily the case, k_{SL} is usually not far from one. And C_{G2} is f_o times the C_{GI} that sets C_{GI}. This k_{SL} is the *self-loading coefficient*.

When f_o is consistent across stages, f_o reduces R_D's by as much as f_o raises C_O's. So t_P's match. And t_P across N stages is N times t_{P1}:

$$t_P = Nt_{P1} = N(69\%R_{D1}C_{O1}) \approx 69\%NR_{D1}C_{GI}(1 + f_o). \quad (32)$$

F_O is very high when C_{GO} is much greater than C_{GI}. With this F_O, the f_o that $F_O^{1/N}$ sets is much greater than one with one stage. f_o falls below this level and approaches one as the number of stages increases.

But as N falls for a fixed F_O, f_o climbs and Nf_o and $N(1 + f_o)$ fall, bottom, and rise. Interestingly, Nf_o bottoms when f_o is e and $N(1 + f_o)$ and t_P bottom when f_o is 3.6. Nf_o/e in Fig. 19 when f_o is e is the optimal N when self-loading (k_{SL}) is negligible. $N(1 + f_o)/(1 + 3.6)$ when f_o is 3.6 nears a more optimal N_0 because k_{SL} is more realistic near one.

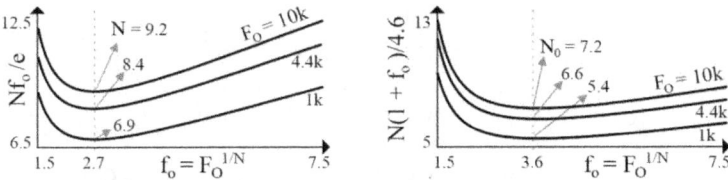

Fig. 19. Optimal gate-driver setting for minimum delay.

M_N in K_1 pulls v_{O1} low when i_N overcomes i_P, after v_{GS} reaches v_T. As v_{GS} rises over v_T and v_{O1} falls across $\Delta v_{O(MAX)}$, M_N enters and remains in triode. M_P is similarly in triode as v_{SG} climbs over v_T and v_O rises. When strengths match, R_{D1} is roughly M_N's triode resistance when v_{GS} nears v_T:

$$R_{D1} \approx R_{N1}\Big|_{v_{DS}<v_{DS(SAT)}}^{v_{GS}>V_{TN0}} \approx \frac{L_{N1} - 2L_{OL}}{W_{N1} K_N'(V_T - V_{TN0})}, \quad (33)$$

where L_{OL} is *overlap length*. C_{GI} is roughly the *channel capacitance* C_{CH1} that W's and L's set across the *oxide capacitance* C_{OX}'' (per unit area):

$$C_{GI} \approx C_{CH1} = C_{CHN1} + C_{CHP1} \approx (W_{N1}L_{N1} + W_{P1}L_{P1})C_{OX}''. \quad (34)$$

Example 7: Determine N and t_P with the inverter from Example 6 when L_{MIN} is 250 nm, L_{OL} is 30 nm, C_{OX}'' is 6.9 fF/μm², and the load is a 100-mm wide, 250-nm long NFET.

Solution:

$W_{N1} = 3$ μm and $W_{P1} = 6.7 W_{N1} = 20$ μm from Example 6

$L_{N1} \equiv L_{P1} \equiv L_{MIN} = 250$ nm

$C_{GI} = C_{GN} + C_{GP} = C_{GN} + 6.7 C_{GN} = 7.7 C_{GN}$

$$F_O = \frac{C_{GO}}{C_{GI}} = \frac{W_{GO}}{7.7 W_{GN}} = \frac{100m}{7.7(3\mu)} = 4.4k$$

$F_O = f_o^{N_0} \equiv 3.6^{N_0} = 4.4k \quad \therefore \quad N \geq N_0 = 6.6 \quad \rightarrow \quad N \equiv 7 \text{ Stages}$

Switched Inductors: Building Blocks

$$R_{D1} \approx \frac{L_{N1} - 2L_{OL}}{W_{N1}K_N'(v_T - V_{TN0})}$$

$$= \frac{250n - 2(30n)}{(3\mu)(200\mu)(2 - 500m)} = 210 \; \Omega$$

$$C_{GI} \approx (W_{N1}L_{N1} + W_{P1}L_{P1})C_{OX}''$$

$$= [(3\mu)(250n) + (20\mu)(250n)](6.9m) = 40 \; fF$$

$$t_P \approx 69\% N R_{D1} C_{GI}(1 + f_o)$$

$$= (69\%)(7)(210)(40f)(1 + 3.6) = 190 \; ps$$

B. Gate-Charge Power

C_{O1} needs *gate charge* q_{G1} to charge across v_O's swing between v_{SS} and v_{DD}. The power supplies deliver this q_{G1} every t_{SW}. So K_1's *gate-charge power* P_{G1} is $q_{G1}(v_{DD} - v_{SS})f_{SW}$, which is equivalent to $C_{O1}(v_{DD} - v_{SS})^2 f_{SW}$:

$$q_{G1} = C_{O1}\Delta v_{O(MAX)} = C_{O1}(v_{DD} - v_{SS}) \tag{35}$$

$$P_{G1} = (v_{DD} - v_{SS})q_{G1}f_{SW} = C_{O1}(v_{DD} - v_{SS})^2 f_{SW}. \tag{36}$$

Of the energy P_{G1} supplies, C_{O1} receives $0.5C_{O1}(v_{DD} - v_{SS})^2$, M_P burns the rest, and M_N later burns C_{O1}'s portion. So K_1 loses all of P_{G1}. Since C_O's climb with f_o, the P_G lost across the driver increases exponentially with N:

$$P_G = \sum_1^N P_{G(K)} = P_{G1} \sum_0^{N-1} f_o^k, \tag{37}$$

where P_{G1} is $P_{G1}f_o^0$ and P_{GN} is $P_{G1}f_o^{N-1}$.

Example 8: Determine P_G for Example 7 when f_{SW} is 1 MHz.
Solution:

$$C_{GI} \approx 40 \; fF \text{ and } N \equiv 7 \text{ from Example 7}$$

$$C_{O1} \approx C_{GI}(1 + f_o) \approx (40f)(1 + 3.6) = 180 \; fF$$

$$P_{G1} = C_{O1}v_{DD}^2 f_{SW} = (180f)(4)^2(1M) = 2.9 \; \mu W$$

$$P_G = P_{G1}\sum_{0}^{N-1} f_o^k \approx (2.9\mu)\sum_{0}^{7-1} 3.6^k = 8.7 \; mW$$

C. Shoot-Through Power

As M_N discharges C_O, M_P conducts *shoot-through current* i_{ST} that M_N sinks. M_N also sinks i_{ST} that M_P supplies when M_P charges C_O. Since this i_{ST} flows from v_{DD} to v_{SS}, v_{DD} and v_{SS} lose *shoot-through power* P_{ST}.

Without C_{O1}, v_I and the *unloaded output* v_{O1}' in Fig. 20 crisscross at v_T. M_N and M_P invert their channels and saturate at this point. Since v_{GS}'s and v_{DS}'s match, i_N and i_P conduct the saturated i_T that v_T over V_{T0} sets:

$$i_T = i_{P/N}\big|_{v_T} = \left(\frac{W_N}{L_N}\right)\left(\frac{K_N'}{2}\right)(v_T - V_{TN0})^2(1 + v_T\lambda_N). \quad (38)$$

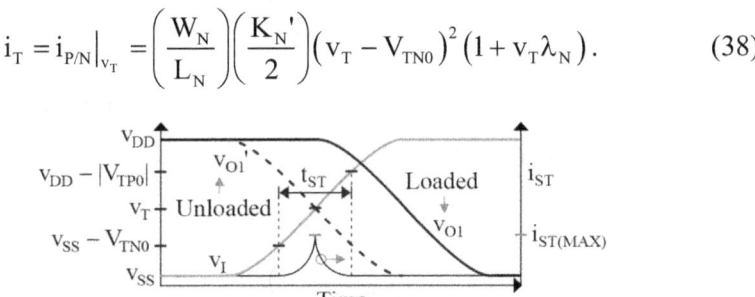

Fig. 20. Inverter's shoot-through response.

i_{ST} is below i_T before v_I reaches v_T because M_N's v_{GS} or M_P's v_{SG} is weaker. i_{ST} is also below i_T after v_I passes v_T because M_P's v_{SG} or M_N's v_{GS} is weaker. So as v_I transitions without C_{O1}, i_{ST} rises to i_T and falls to zero.

But since C_{O1} delays v_{O1}, v_I reaches v_T before v_{O1} does. So at v_T, M_N's v_{DS} or M_P's v_{SD} is lower than the other's. As a result, i_P sets i_{ST} with a lower v_{SD} when v_{O1} falls and i_N sets i_{ST} with a lower v_{DS} when v_{O1} rises. In other words, i_{ST} does not reach i_T.

When f_o is one (C_{G2} equals C_{G1}), i_{ST} can reach 15% of i_T. $i_{ST(MAX)}$ falls with higher C_O's below this point. So across transitions, i_{ST} rises and peaks at about $30\% i_T/(1 + f_o)$ before falling back to zero.

M_N and M_P conduct i_{ST} when v_I is between MOS thresholds: V_{TN0} above v_{SS} and $|V_{TP0}|$ below v_{DD}. The v_O that sets this v_I is from another stage whose R_D and C_O resemble R_{D1} and C_{O1}. So *shoot-through time* t_{ST} is the time v_I needs to traverse between MOS thresholds:

$$t_{ST} = t_{RC}\Big|_{v_{SS}+V_{TN0}}^{v_{DD}-|V_{TP0}|} \approx \tau_{RC} \ln\left[\frac{(v_{DD}-v_{SS})-(v_{DD}-|V_{TP0}|)}{(v_{DD}-v_{SS})-(v_{SS}+V_{TN0})}\right]^{-1}. \quad (39)$$

i_{ST} averages about one-third of $i_{ST(MAX)}$ because i_{ST} scales quadratically with the v_{GS} that v_I sets and v_I scales linearly with time. P_{ST1} is therefore the power v_{DD} and v_{SS} burn with i_{ST1}'s average $i_{ST1(MAX)}/3$ across two t_{ST} fractions of t_{SW}, once when v_O rises and again when v_O falls:

$$P_{ST1} = i_{ST1}(v_{DD}-v_{SS})\left(\frac{2t_{ST}}{t_{SW}}\right) \approx \left[\frac{30\% i_{T1}}{3(1+f_o)}\right](v_{DD}-v_{SS})\left(\frac{2t_{ST}}{t_{SW}}\right). \quad (40)$$

P_{ST}'s scale with f_o^k because i_T's are f_o times higher than the preceding stage's i_T. So the P_{ST} lost across the driver climbs exponentially with N:

$$P_{ST} = \sum_1^N P_{ST(K)} \approx P_{ST1}\sum_0^{N-1} f_o^k, \quad (41)$$

where P_{ST1} is $P_{ST1}f_o^0$ and P_{STN} is $P_{ST1}f_o^{N-1}$.

Example 9: Determine P_{ST} for Example 7 when λ_N is 5%.
Solution:

$$v_T = 2\text{ V}, W_N = 3\text{ μm}, L_N = 250\text{ nm}, R_{D1} = 210\text{ Ω},$$
$$\text{and } C_{GI} = 40\text{ fF from Example 7}$$

$$i_{T1} = \left(\frac{W_N}{L_N - 2L_{OL}}\right)\left(\frac{K_N'}{2}\right)(v_T - V_{TN0})^2(1+v_T\lambda_N)$$

$$= \left(\frac{3\mu}{190n}\right)\left(\frac{200\mu}{2}\right)(2-500m)^2[1+(2)5\%] = 4.0\text{ mA}$$

$$\tau_{RC} \approx R_{D1}C_{GI}(1+f_o) = (210)(40f)(1+3.6) = 39\text{ ps}$$

$$t_{ST} \approx \tau_{RC} \ln \frac{V_{DD} - V_{TN0}}{|V_{TP0}|} = (39p) \ln \frac{4 - 500m}{700m} = 63 \text{ ps}$$

$$P_{ST1} \approx \left[\frac{30\% i_{T1}}{2(1+f_o)} \right] V_{DD} \left(\frac{2t_{ST}}{t_{SW}} \right)$$

$$= \left[\frac{30\%(4.0m)}{3(1+3.6)} \right] (4) \left[\frac{2(63p)}{1\mu} \right] = 44 \text{ nW}$$

$$P_{ST} \approx P_{ST1} \sum_0^{N-1} f_o^k \approx (44n) \sum_0^{7-1} 3.6^k = 130 \text{ }\mu W$$

Note: P_{ST} is 1.5% of P_G.

D. Design Notes

This analysis shows t_P is minimal when f_o nears 3.6, t_P scales with N, and P_G and P_{ST} climb exponentially with N. With f_o fixed, F_O sets N. And since F_O is usually high in power supplies, this N is also high. So t_P, P_G, and P_{ST} are that much higher than for the lightly loaded minimum-size inverter.

P_{ST} maxes without C_O. P_{ST} is roughly 15% of this maximum when f_o is one and 6.5% when f_o is 3.6. At this level, P_{ST} is a small fraction of P_G. So total *gate-driver power* P_{DRV} is mostly the P_G that C_O's set.

Delay scales with $N(1 + f_o)$ and power with Σf_o^k up to f^{N-1}. So P_{DRV} is more sensitive to N than t_P is. In this light, reducing N favors P_{DRV} savings over t_P. In power supplies, limiting N to three or five and setting f_o with this N to a level that is higher than 3.6 is fairly routine for this reason.

In practice, R_{DI} and C_{GI} are dynamic parameters. Engineers often skew W's in the last stage to minimize the i_{DS}–v_{DS} *overlap power* P_{IV} lost across the power switch. And power switches are sometimes off chip. Even with these variations, fewer stages usually save more power than extend delay.

Example 10: Determine t_P and P_{DRV} for Example 7 when N is 4.

Solution:

$F_O = 4.4k$, $R_{DI} = 210\ \Omega$, and $C_{GI} = 40$ fF from Example 7

$F_O = f_o^N \equiv f_o^4 = 4.4k\quad \therefore\quad f_o = 8.1$

$t_P \approx 69\% N R_{DI} C_{GI} (1 + f_o)$

$\quad = (69\%)(4)(210)(40f)(1 + 8.1) = 210$ ps

$C_{OI} \approx C_{GI}(1 + f_o) \approx (40f)(1 + 8.1) = 360$ fF

$P_{GI} = C_{OI} V_{DD}^2 f_{SW} = (360f)(4)^2(1M) = 5.8\ \mu W$

$P_G \approx P_{GI} \sum_0^{N-1} f_o^k \approx (5.8\mu) \sum_0^{4-1} 8.1^k = 3.5$ mW

$\tau_{RC} \approx R_{DI} C_{GI}(1 + f_o) = (210)(40f)(1 + 8.1) = 76$ ps

$t_{ST} \approx \tau_{RC} \ln\left(\dfrac{V_{DD} - V_{TN0}}{|V_{TP0}|}\right) = (76p)\ln\left(\dfrac{4 - 500m}{700m}\right) = 120$ ps

$P_{ST1} \approx \left[\dfrac{30\% i_{T1}}{3(1 + f_o)}\right] V_{DD} \left(\dfrac{2 t_{ST}}{t_{SW}}\right)$

$\quad = \left[\dfrac{30\%(4.0m)}{3(1 + 8.1)}\right](4)\left[\dfrac{2(120p)}{1\mu}\right] = 42$ nW

$P_{ST} \approx P_{ST1} \sum_0^{N-1} f_o^k \approx (42n) \sum_0^{4-1} 8.1^k = 26\ \mu W$

$P_{DRV} = P_G + P_{ST} = 3.5m + 22\mu = 3.5$ mW

Note: Reducing N to 4 extends t_P 11% and reduces P_{DRV} 40%.

3.4. Dead-Time Logic

Power switches that connect to the same switching node should not close at the same time. If they do, they would short v_{IN} or v_O to ground. Even if the switches do not melt, they would burn too much power.

Engineers insert *dead time* t_{DT} between adjacent switches for this reason. This way, even if gate signals overlap, switches cannot close at the same time. Consider gate command v_G' and *gate voltages* v_G and v_{GX} in Fig. 21, for example. v_G rises a t_{DT} after adjacent v_{GX} falls, even when v_G' commands a high before v_{GX} falls.

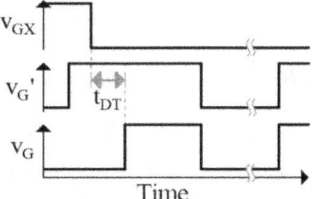

Fig. 21. Dead-time response.

t_{DT} only applies to closing events, when burning unnecessary power is possible. Opening switches is inherently safe because it stops current flow. This is why v_G falls with v_G' without delay in Fig. 21.

t_{DT} should delay turn-on commands only when adjacent commands are high, when shorting events are possible. Whenever safe, reducing delays helps the system respond and recover more quickly. This is why v_G in Fig. 21 rises with v_G' without delay the second time v_G' transitions high.

To assert these states, *dead-time circuits* should sense, delay, and compare neighboring commands with v_G'. R's and C's in Fig. 22, for example, sense and delay neighboring v_G's. And NOR and NAND gates compare these delayed signals with v_G' to determine v_G. R's and C's or inverter chains can sense and delay neighboring v_G's.

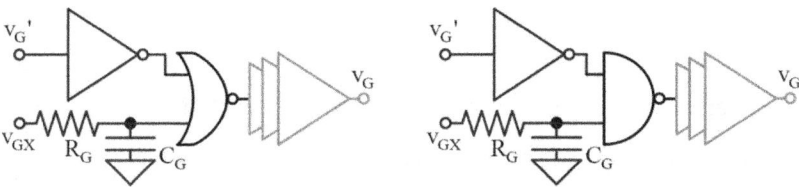

Fig. 22. Active-high and -low dead-time circuits.

The NOR gate outputs an active-high command when v_G' is high after neighboring v_G's fall. The NAND gate outputs an active-low command when v_G' is low after neighboring v_G's rise. The NOR gate deactivates v_G as soon as v_G' falls and the NAND gate deactivates v_G as soon as v_G' rises.

4. Comparator Blocks

4.1. Comparators

A *comparator* is an *analog–digital converter* (ADC) that compares two analog voltages and outputs a digital voltage to indicate which is higher. v_O is high when the *positive input* v_P overcomes the *negative input* v_N and low when v_N overcomes v_P. In other words, v_O is high when the *differential input voltage* v_{ID} between v_P and v_N is positive and low when v_{ID} is negative. So comparators are also *polarity detectors*.

For this functionality, comparators usually sense and amplify v_{ID} until v_O saturates near v_{DD} or v_{SS}. In this respect, comparators are like op amps. The only difference is comparators do not need stabilizing components.

So like op amps, most comparators sense v_{ID} in Fig. 23 with a differential transconductor G_D and amplify with resistance R_A. This way, the low-frequency differential voltage gain A_{D0} from v_{ID} to v_A is $G_D R_A$. And with enough A_{D0}, small v_{ID}'s can swing v_A towards the supplies.

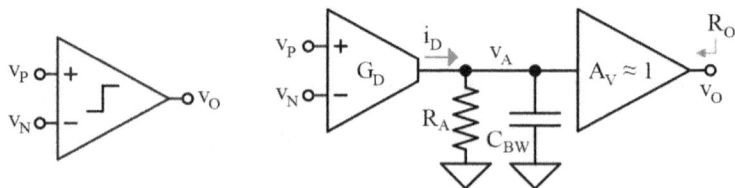

Fig. 23. Typical comparator.

C_{BW} embeds the unintended capacitances that shunt R_A past the pole that sets t_P. The voltage buffer A_V sets the output resistance R_O with which v_O sets a state. Note G_D and A_V cannot swing v_A or v_O past v_{SS} or v_{DD}.

4.2. Hysteretic Comparators

Hysteretic comparators trip high when v_{ID} rises over the *rising threshold* $v_{T(HI)}$ and low when v_{ID} falls under the *falling threshold* $v_{T(LO)}$. Between these thresholds, v_O holds the previous state. The difference between these thresholds is the comparator's *hysteresis* Δv_T.

In Fig. 24, the top and bottom comparators set and reset the flip flop when v_{ID} rises over $v_{T(HI)}$ and falls under $v_{T(LO)}$. Between thresholds, the flip flop holds the previous state. Since v_{ID} cannot rise over $v_{T(HI)}$ and fall under $v_{T(LO)}$ at the same time, set/reset dominance is irrelevant.

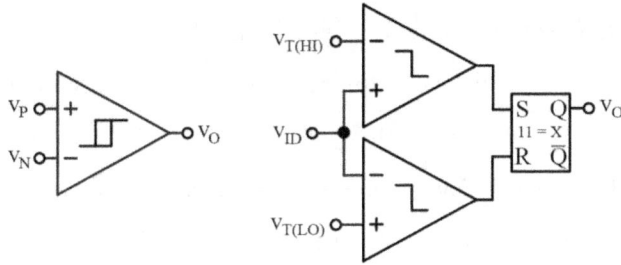

Fig. 24. Flip-flopped hysteretic comparator.

Some implementations use positive feedback to latch the state of the comparator past its trip point. Holding the state this way effectively splits the trip point apart into separate thresholds. So like in digital flip flops, positive feedback holds the previous state. The effect of this feedback on comparators establishes hysteresis.

4.3. Summing Comparators

Summing comparators add their inputs. This way, v_O trips high when v_P's overcome v_N's and low when v_N's surpass v_P's. So v_O approaches v_{DD} when the *differential sum* Σv_{ID} is positive and v_{SS} when Σv_{ID} is negative.

Paralleling differential transconductors into a comparator is one way of adding inputs. G_{D1} and G_{D2} in Fig. 25, for example, add i_{D1} and i_{D2}, R_S drops Σv_D, and CP_O compares Σv_D to zero. When G_D's match, i_{D1} and i_{D2} are $\Sigma v_{ID} G_D$ and Σv_D is $\Sigma v_{ID} G_D R_S$. This way, Σv_{ID}'s polarity matches Σv_D's. So CP_O trips v_O high and low when Σv_{ID} and Σv_D are positive and negative:

$$\Sigma v_D = (i_{D1} + i_{D2})R_S = \left[(v_{P1} - v_{N1}) + (v_{P2} - v_{N2})\right]G_D R_S = \Sigma v_{ID} G_D R_S. \quad (42)$$

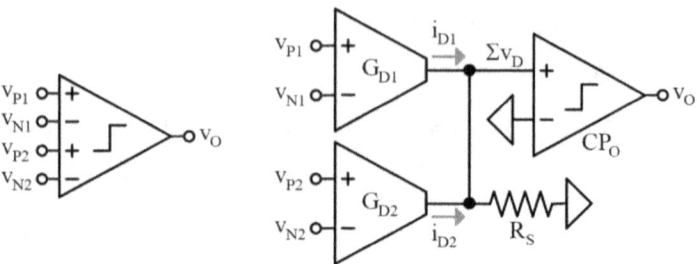

Fig. 25. Summing comparator.

Summing hysteretic comparators add inputs the same way. In Fig. 26, for example, G_{D1} and G_{D2} add i_{D1} and i_{D2} into R_S and a hysteretic comparator compares the resulting Σv_D to zero. When G_D's match $1/R_S$, the $\Sigma v_{ID}G_D R_S$ that sets Σv_D reproduces Σv_{ID}. So in effect, v_O trips when Σv_{ID} climbs over or below CP_Σ's $v_{T(HI)}$ or $v_{T(LO)}$. Other implementations integrate CP_Σ's v_T and feedback mechanics together with G_{D1} and G_{D2}.

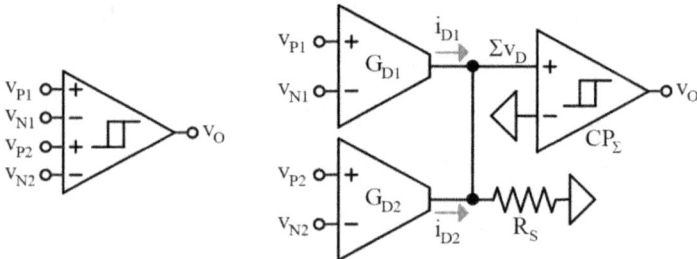

Fig. 26. Summing hysteretic comparator.

5. Timing Blocks

5.1. Clocked Sawtooth Generator

A *clocked sawtooth generator* outputs a *sawtooth voltage* v_S that ramps across every *clock period* t_{CLK}. This type of circuit needs a clock input v_{CLK}, a ramp generator, and a reset circuit. In Fig. 27, I_S and C_S ramp v_S and the inverter chain, SR flip flop, and M_R reset v_S.

The inverters output a delayed version v_D of v_{CLK}. v_{CLK} sets the flip flop to start a reset event and the inverters end the "reset" when v_D resets the flip flop. This way, the flip flop pulses v_R across the *inverters' delay* t_I.

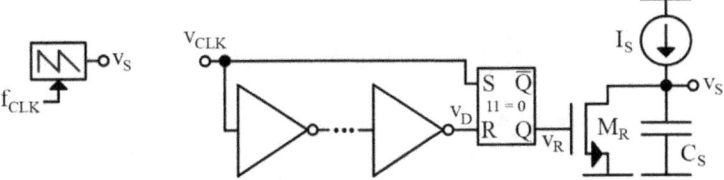

Fig. 27. Clocked sawtooth generator.

M_R should discharge C_S before t_I elapses. Since v_R reaches v_{DD} when M_R resets v_S and v_{DD} is typically much higher than $v_{S(HI)}$, the v_{GS} that v_R sets usually overwhelms the v_{DS} that $v_{S(HI)}$ sets. So M_R's triode resistance R_R collapses most of v_S. This *reset time* t_R should be shorter than t_I:

$$t_R \approx t_{90\%\Delta v_{O(MAX)}} = \tau_{RC} \ln(1-90\%)^{-1} = 2.3\tau_{RC} = 2.3 R_R C_S. \quad (43)$$

v_R falls quickly across v_{DD} and v_{SS} after t_I elapses. This transition is so fast that M_R's C_{GD} injects noise into C_S. The voltage-divided noise that C_{GD} drops across C_S shifts $v_{S(LO)}$ below v_{SS}. Since M_R is in triode when this happens, C_{GD} is roughly half of C_{CH}, which is usually much lower than C_S:

$$v_{S(LO)} \approx v_{SS} - \Delta v_R \left(\frac{C_{GD}}{C_{GD}+C_S}\right) \approx v_{SS} - (v_{DD}-v_{SS})\left(\frac{0.5 C_{CH}}{C_S}\right). \quad (44)$$

After v_R falls, I_S ramps v_S from this $v_{S(LO)}$ to $v_{S(HI)}$ across the part of t_{CLK} that excludes t_I:

$$v_{S(HI)} \approx v_{S(LO)} + \left(\frac{I_S}{C_S}\right)(t_{CLK}-t_I). \quad (45)$$

Example 11: Determine I_S and W_R so $v_{S(HI)}$ is 300 mV when v_{DD} is 1 V, C_S is 5 pF, t_{CLK} is 1 μs, t_I is 2 ns, V_{TN0} is 500 mV, K_N' is 200 μA/V², C_{OX}'' is 6.9 fF/μm², L_R is 250 nm, and L_{OL} is 30 nm.

Solution:

$t_R \approx 2.3 R_R C_S = 2.3 R_R(5p) \equiv 25\% t_I = 25\%(2n) = 500$ ps

∴ $R_R \leq 44\ \Omega$

$$R_R \approx \frac{L_R - 2L_{OL}}{W_R K_N'(V_{DD} - V_{TN0})} = \frac{250n - 2(30n)}{W_R(200\mu)(1 - 500m)} \leq 44\ \Omega$$

$$\therefore\quad W_R \geq 43\ \text{nm} \quad \rightarrow \quad W_R \equiv 43\ \mu\text{m}$$

$$V_{S(LO)} \approx -V_{DD}\left(\frac{0.5 W_R L_R C_{OX}''}{2C_S}\right)$$

$$= -(1)\left[\frac{(0.5)(43\mu)(250n)(6.9m)}{5p}\right] = -7\ \text{mV}$$

$$V_{S(HI)} \approx V_{S(LO)} + \left(\frac{I_S}{C_S}\right)(t_{CLK} - t_I)$$

$$\approx -7m + \left(\frac{I_S}{5p}\right)(1\mu - 2n) \equiv 300\ \text{mV} \quad \therefore\quad I_S = 1.5\ \mu\text{A}$$

5.2. Sawtooth Oscillator

A *sawtooth oscillator* does not need a clock to generate v_S. In fact, it is a clock. And in the case of Fig. 28, it is also a *relaxation oscillator*.

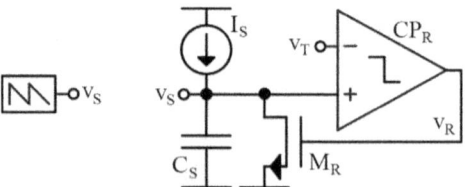

Fig. 28. Sawtooth oscillator.

Here, I_S and C_S ramp v_S and comparator CP_R resets v_S after v_S overcomes v_T. v_R sets $v_{S(LO)}$ when v_R falls. v_R trips after v_S climbs over v_T across CP_R's rising t_P^+ to $v_{S(HI)}$. Since v_S falls below v_T when M_R resets v_S, M_R should collapse v_S before CP_R's falling t_P^- elapses. This way, $v_{S(HI)}$ is

$$V_{S(HI)} = V_T + \left(\frac{I_S}{C_S}\right)t_P^+, \tag{46}$$

where M_R's t_R is shorter than t_P^-.

t_{CLK} is the time C_S and CP_R need to ramp v_S from $v_{S(LO)}$ to $v_{S(HI)}$ (which is t_P^+ after v_S reaches v_T) and the time CP_R needs to start another cycle:

$$t_{CLK} = \left(\frac{C_S}{I_S}\right)(v_{S(HI)} - v_{S(LO)}) + t_P^- = \left(\frac{C_S}{I_S}\right)(v_T - v_{S(LO)}) + t_P^+ + t_P^-. \quad (47)$$

Since I_S, C_S, and t_P vary with temperature and fabrication runs, t_{CLK}'s tolerance is usually fairly high. Trimming and balancing variations across temperature reduce this, but only with more silicon area and test time.

Example 12: Determine v_T, I_S, and W_R so $v_{S(HI)}$ is 300 mV and t_{CLK} is 1 μs when v_{DD} is 1 V, C_S is 5 pF, t_P is 100 ns, V_{TN0} is 500 mV, K_N' is 200 μA/V², C_{OX}'' is 6.9 fF/μm², L_R is 250 nm, L_{OL} is 30 nm, and W_{MIN} is 3 μm.

Solution:

$$t_R \approx 2.3 R_R C_S = 2.3 R_R (5p) \equiv 25\% t_P = 25\%(100n) = 25 \text{ ns}$$

$$\therefore R_R \leq 2.2 \text{ k}\Omega$$

$$R_R \approx \frac{L_R - 2L_{OL}}{W_R K_N'(V_{DD} - V_{TN0})} = \frac{250n - 2(30n)}{W_R(200\mu)(1 - 500m)} \leq 2.2 \text{ k}\Omega$$

$$\therefore W_R \geq 860 \text{ nm} \quad \rightarrow \quad W_R \equiv 3 \text{ μm}$$

$$v_{S(LO)} \approx -V_{DD}\left(\frac{0.5 W_R L_R C_{OX}''}{2C_S}\right)$$

$$= -(1)\left[\frac{(0.5)(3\mu)(250n)(6.9m)}{5p}\right] = -520 \text{ μV}$$

$$t_{CLK} = \left(\frac{C_S}{I_S}\right)(v_{S(HI)} - v_{S(LO)}) + t_P^-$$

$$= \left(\frac{5p}{I_S}\right)[300m - (-520\mu)] + 100n \equiv 1 \text{ μs}$$

$$\therefore I_S = 1.7 \text{ μA}$$

$$V_{S(HI)} = V_T + \left(\frac{I_S}{C_S}\right)t_P^+ = V_T + \left(\frac{1.7\mu}{5p}\right)(100n) \equiv 300 \text{ mV}$$

$$\therefore \quad V_T = 270 \text{ mV}$$

5.3. One-Shot Oscillator

One-shot oscillators are *pulse generators*. But more specifically, they are interruptible relaxation oscillators. They pulse once when pulsed and pulse continuously when kept enabled.

Like *ring oscillators*, they usually close inverting feedback loops that internal components delay. In the case of Fig. 29, CP_R, the flip flop, and M_R close the loop and I_S and C_S delay it. v_O pulses each time v_I pulse-sets the flip flop and oscillates between states when v_I stays high.

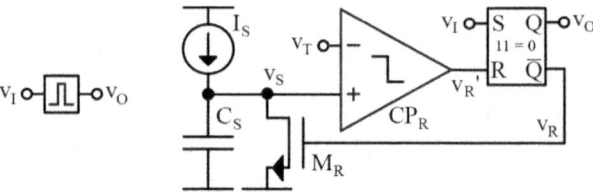

Fig. 29. One-shot oscillator.

Raising v_I sets v_O and opens M_R to start a pulse. I_S charges C_S across the pulse until M_R resets v_S. The *pulse width* t_{PW} is the time C_S, CP_R, and the reset-dominant flip flop need to ramp v_S to v_T, trip, and reset:

$$t_{PW} = \left(\frac{C_S}{I_S}\right)\left(V_{S(HI)} - V_{S(LO)}\right) = \left(\frac{C_S}{I_S}\right)\left(V_T - V_{S(LO)}\right) + t_P^+ + t_{SR}^+. \quad (48)$$

So across this t_{PW}, I_S charges C_S over v_T across t_P^+ and t_{SR}^+:

$$V_{S(HI)} = V_T + \left(\frac{I_S}{C_S}\right)\left(t_P^+ + t_{SR}^+\right). \quad (49)$$

Keeping v_I high sets another pulse when the reset command v_R' falls. This happens after v_S falls below v_T. The time between pulses t_{OFF} is the time CP_R and the flip flop need to trip v_R' and set v_O. So the *oscillation period* t_{OSC} is

$$t_{OSC} = t_{PW} + t_{OFF} \approx t_{PW} + t_P^- + t_{SR}^-. \qquad (50)$$

To collapse v_S before t_{OFF} elapses, M_R's t_R should be shorter than t_{OFF}.

Example 13: Determine v_T, I_S, W_R, and t_{OSC} so $v_{S(HI)}$ is 300 mV and t_{PW} is 750 ns when v_{DD} is 1 V, C_S is 5 pF, t_P is 100 ns, t_{SR} is 1 ns, V_{TN0} is 500 mV, K_N' is 200 µA/V², C_{OX}'' is 6.9 fF/µm², L_R is 250 nm, L_{OL} is 30 nm, and W_{MIN} is 3 µm.

Solution:

$$t_{OFF} = t_P^- + t_{SR}^- = 100n + 1n = 100 \text{ ns}$$

$$t_{OSC} = t_{PW} + t_{OFF} = 750n + 101n = 850 \text{ ns}$$

$$t_R \approx 2.3 R_R C_S = 2.3 R_R (5p) \equiv 25\% t_{OFF} = 25\%(100n) = 25 \text{ ns}$$

$\rightarrow$ Same as Example 12 $\therefore$ $W_R \equiv 3$ µm, $v_{S(LO)} = -520$ µV

$$t_{PW} = \left(\frac{C_S}{I_S}\right)\left(v_{S(HI)} - v_{S(LO)}\right) = \left(\frac{5p}{I_S}\right)(300m + 520\mu) \equiv 750 \text{ ns}$$

$\therefore$ $I_S = 2.0$ µA

$$v_{S(HI)} = v_T + \left(\frac{I_S}{C_S}\right)\left(t_P^+ + t_{SR}^+\right)$$

$$= v_T + \left(\frac{2.0\mu}{5p}\right)(100n + 1n) \equiv 300 \text{ mV}$$

$\therefore$ $v_T = 260$ mV

---

6. Switch Blocks

6.1. Class-A Inverters

Inverters are comparators that compare v_I to a v_P that v_T sets. In the case of *class-A inverters*, v_O trips when v_I climbs over or below the v_T that M_I's

v_{GS} or v_{SG} in Fig. 30 sets with v_{SS} or v_{DD}. These inverters are useful when transitioning analog v_{GS}-level swings to digital supply-level swings.

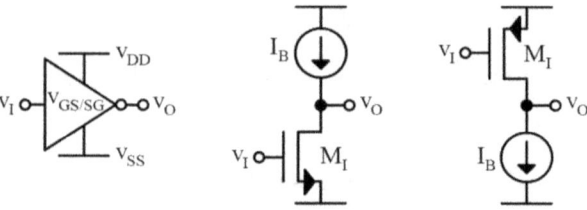

Fig. 30. Class-A inverter.

These circuits balance when v_I is the v_G needed to sustain I_B with v_O halfway between the supplies. M_I or I_B pulls v_O low when v_I is higher and I_B or M_I pulls v_O high when v_I is lower. This balancing v_G, which sets v_T, is V_{GS} over v_{SS} or V_{SG} below v_{DD} when M_I carries I_B in saturation:

$$v_T = V_{GS}\Big|_{I_B}^{V_{DS} > V_{DS(SAT)}} + v_{DD/SS}. \tag{51}$$

Example 14: Determine W and L so v_I is 650 mV when v_{DD} is 4 V, I_B is 10 μA, W_{MIN} is 3 μm, L_{OL} is 30 nm, K_N' is 200 μA/V², V_{TN0} is 500 mV, and λ_N is 5%.

Solution:

$$v_I \equiv 650 \text{ mV} > V_{TN0} = 500 \text{ mV} \quad \therefore \quad \text{Inversion}$$

$$V_{DS} = 0.5 v_{DD} = 0.5(4) = 2 \text{ V}$$

$$I_B = \left(\frac{W_{CH}}{L_{CH}}\right)\left(\frac{K_N'}{2}\right)(v_T - V_{TN0})^2(1 + \lambda_N V_{DS})$$

$$= \left(\frac{W_{CH}}{L_{CH}}\right)\left(\frac{200\mu}{2}\right)(650m - 500m)^2[1 + 5\%(2)] = 10 \text{ μA}$$

$$\therefore \quad \frac{W_{CH}}{L_{CH}} = 4.0$$

$$\rightarrow \quad W = W_{CH} \equiv W_{MIN} = 3 \text{ μm}$$

$$L_{CH} = L - 2L_{OL} = L - 2(30n) \equiv \frac{W_{CH}}{4.0} = \frac{3\mu}{4.0} = 750 \text{ nm}$$

$$L = 810 \text{ nm}$$

6.2. Supply-Sensing Comparators

A. Low Side

Low-side comparators sense and compare v_{SS}-level voltages. In the case of Fig. 31, v_{GS}'s for M_L, M_N, and M_B match when the *low input* v_L equals v_N. So M_L, M_N, and M_B sink I_B and M_{M2} mirrors the I_B that M_N pulls from M_{M1}. With all currents matched, the circuit balances.

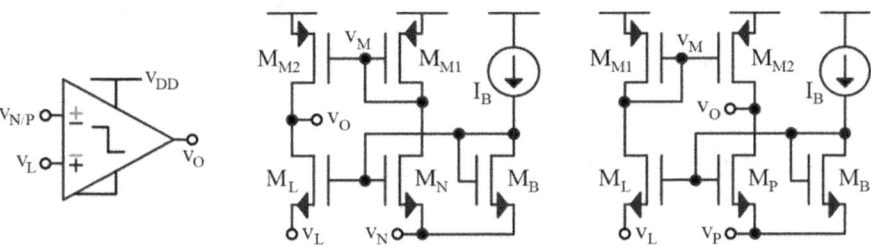

Fig. 31. Positive and negative low-side comparators.

When v_L climbs over v_N, M_L's v_{GS} shrinks. This reduces M_L's current, allowing M_{M2}'s I_B to charge capacitances at v_O towards v_{DD}. The operation reverses when v_L falls under v_N: M_L's v_{GS} grows, so M_L pulls v_O low.

M_B connects to v_N to "crush" I_B when v_N nears v_{DD}. A high enough v_N shuts off M_B, M_N, and M_{M1} and raises M_L's v_{GS}, so M_L pulls v_O low towards v_L. I_B re-activates M_B, M_N, and M_{M1} automatically when v_N falls.

B. High Side

High-side comparators sense and compare v_{DD}-level voltages. In the case of Fig. 32, v_{SG}'s for M_H, M_N, and M_B match when the *high input* v_H equals v_N. So M_H, M_N, and M_B conduct I_B and M_{M2} mirrors the I_B that M_N feeds M_{M1}. With all currents matched, the circuit balances.

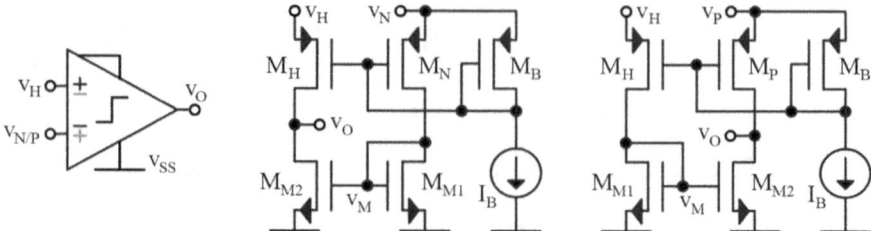

Fig. 32. Positive and negative high-side comparators.

When v_H climbs over v_N, M_H's v_{SG} grows. This raises M_H's current i_H, charging capacitances at v_O towards v_H. The operation reverses when v_H falls under v_N: M_H's v_{SG} shrinks and i_H falls, so M_{M2}'s I_B pulls v_O low.

M_B connects to v_N to "crush" I_B when v_N nears v_{SS}. A low enough v_N shuts off M_B, M_N, and M_{M1} and raises M_I's v_{SG}, so M_I pulls v_O high towards v_I. I_B re-activates M_B, M_N, and M_{M1} automatically when v_N rises.

C. Offset

Integrating a *systemic offset* $V_{OS(S)}$ into the comparator is often useful. This way, v_O trips when $v_{L/H}$ climbs $V_{OS(S)}$ over or falls under v_N. Favoring one input over the other this way amounts to reducing its current density: widening its transistor or reducing its current. The resulting $V_{OS(S)}$ is the difference of v_{GS}'s when the circuit balances.

$V_{OS(S)}$ is the difference of logarithms in sub-threshold and saturation voltages in inversion. So $V_{OS(S)}$ is the ratio or square-root difference of current densities. When currents match I_B, widening $M_{L/H}$'s $W_{L/H}$ sets a $V_{OS(S)}$ that favors $v_{L/H}$ and trips v_O high when $v_{L/H}$ climbs $V_{OS(S)}$ over v_N:

$$V_{OS(S)}\Big|_{V_{DS}>3V_t}^{V_{GS}<v_T} = V_{GSL/H} - V_{GSN} \approx n_I V_t \ln\left[\frac{I_{L/H}(W/L)_N}{(W/L)_{L/H} I_N}\right] \quad (52)$$

$$V_{OS(S)}\Big|_{V_{DS}>V_{GST}}^{V_{GS}>v_T} = V_{GSL/H} - V_{GSN} \approx \sqrt{\frac{2I_{L/H}}{K'(W/L)_{L/H}}} - \sqrt{\frac{2I_N}{K'(W/L)_N}}. \quad (53)$$

Since these comparators balance when v_O matches v_M, $M_{L/H}$ and M_N should saturate (by design) when v_O equals v_M.

Example 15: Determine W_L so $V_{OS(S)}$ favors v_L with 10 mV when I_B is 10 µA, W_N is 3 µm, L's are 1 µm, L_{OL} is 30 nm, and K_N' is 200 µA/V².

Solution:

$$L_{CH} = L_L - 2L_{OL} = 1\mu - 2(30n) = 940 \text{ nm}$$

$$V_{OS(S)} \approx \sqrt{\frac{2I_B L_{CH}}{K_N'}} \left(\sqrt{\frac{1}{W_L}} - \sqrt{\frac{1}{W_N}} \right)$$

$$= \sqrt{\frac{2(10\mu)(940n)}{200\mu}} \left(\sqrt{\frac{1}{W_L}} - \sqrt{\frac{1}{3\mu}} \right) \equiv 10 \text{ mV}$$

$\therefore \quad W_L = 2.7 \text{ µm}$

D. Design Notes

These comparators are compact, fast, and low power. With only two nodes and four transistors in its core, silicon area, stray capacitances, and power consumption are low. The circuit also shuts down and reactivates automatically. This on-demand feature saves quiescent power.

In practice, v_{ID} can require 5–15 mV to trip the comparator. Cascading an inverter can reduce this five to ten times. But since the circuit trips when it balances, v_O's tripping point equals the v_M that M_{M1}'s v_{GS} sets with I_B and v_{SS} or v_{DD}. So v_O should feed a class-A inverter with a similarly v_{GS}-set threshold.

$v_{L/H}$ need not always be the positive input. Cascading an inverter is one way of inverting the polarity. Flipping the mirroring transistors M_{M1} and M_{M2} is another. With this last method, M_L or M_H feeds M_{M1}, M_P and M_{M2} set v_O, and v_O trips low or high when v_L or v_H rises over or falls under v_P.

6.3. Zero-Current Detectors

Systems enter *discontinuous-conduction mode* (DCM) when *zero-current detectors* (ZCD) determine L_X's drain current $i_{L(D)}$ reaches zero. ZCD's normally monitor the voltage across a drain switch for this purpose. Since its v_{DS} scales with $i_{L(D)}$, $i_{L(D)}$ crosses zero when v_{DS} crosses zero.

Buck-based supplies use a *low-side switch* M_L to drain L_X. $i_{L(D)}$ flows from v_L towards the v_{SW} that connects to L_X. So v_L in Fig. 33 is usually higher than v_{SW}. This prompts the low-side comparator CP_{ZL} to trip the class-A inverter low. v_O trips high when $i_{L(D)}$ reverses v_{ID}'s polarity.

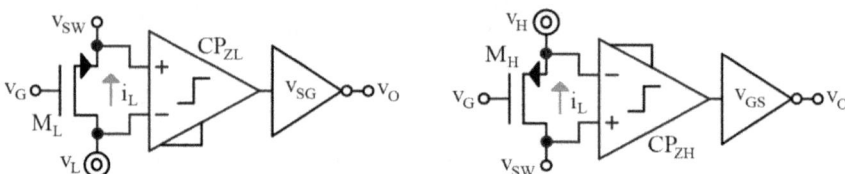

Fig. 33. Low- and high-side zero-current detectors.

Boost-based supplies use a *high-side switch* M_H to drain L_X. $i_{L(D)}$ flows from the v_{SW} that connects to L_X towards v_H. So v_{SW} is usually higher than v_H in Fig. 33. This prompts the high-side comparator CP_{ZH} to trip the class-A inverter low. v_O trips high when $i_{L(D)}$ reverses v_{ID}'s polarity.

In practice, ZCD's trip t_P after $i_{L(D)}$ reverses. Some engineers add an offset to CP_Z that favors $v_{L/H}$ to compensate for this delay. This way, the ZCD detects when $v_{L/H}$ is within $V_{OS(S)}$ of v_{SW} and trips before $v_{L/H}$ reaches v_{SW}. Although not perfect nor always necessary, anticipating the transition can keep $i_{L(D)}$ from reversing and consuming unnecessary power.

6.4. Ring Suppressor

When the ZCD opens the switches in DCM, v_{SWI} in Fig. 34 nears ground and v_{SWO} is close to v_O. This voltage across L_X induces an i_L that charges and discharges the capacitances C_{SWI} and C_{SWO} at v_{SWI} and v_{SWO}. L_X and C_{SW}'s exchange energy E_{LC} at their *LC frequency* f_{LC} until L_X's R_L burns it.

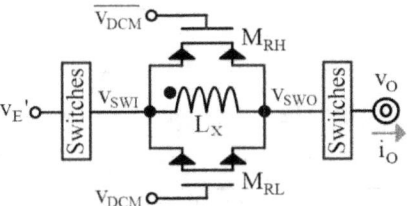

Fig. 34. Ring suppressor.

As R_L dampens oscillations, v_{SWI} and v_{SWO} approach one another until v_L is zero. Since bucks and boosts exclude input or output switches, v_{SW} approaches v_O in bucks and v_{IN} in boosts. The C_{SW} that sets the *LC time constant* τ_{LC} in buck–boosts is the series combination of C_{SWI} and C_{SWO}.

These oscillations can last several cycles. This is unfortunate because the *electromagnetic interference* (EMI) i_L generates can alter the feedback action that sets v_O or i_O. The purpose of $M_{RL/H}$ is to suppress this "ringing".

The *ring suppressor* is a resistor that the ZCD invokes when L_X enters DCM. This R_R should burn E_{LC} before L_X receives it. Since C_{SW} and L_X exchange E_{LC} every quarter *LC period* t_{LC}, t_{RC} should be less than 25% t_{LC}:

$$t_S = t_{RC} \approx 2.3\tau_{RC} = 2.3 R_R C_{SW} < \frac{t_{LC}}{4} = \frac{2\pi\tau_{LC}}{4} = \frac{2\pi/4}{\sqrt{L_X C_{SW}}}, \quad (54)$$

where *suppression time* t_S is roughly the t_{RC} needed to collapse 90% of v_L.

Low- and high-side switches can realize this R_R. A PFET M_{RH} may be sufficient when v_O is high and an NFET M_{RL} when v_{IN} exceeds v_O and v_O is low. Both may be necessary when v_{IN} is not much greater than a low v_O. Regardless, these FETs are often mostly in triode when they collapse v_L.

Example 16: Determine W_R when v_O for a buck is 1.8 V, L_X is 10 μH, C_{SW} is 5 pF, W_{MIN} is 3 μm, L_R is 250 nm, L_{OL} is 30 nm, K_{P}' is 40 μA/V^2, and V_{TP0} is –700 mV.

Solution:

$$L_{CH} = L_R - 2L_{OL} = 250n - 2(30n) = 190 \text{ nm}$$

$$t_S \approx 2.3 R_R C_{SW} = 2.3 R_R (5p) < \frac{\pi}{2}\sqrt{L_X C_{SW}} = \frac{\pi}{2}\sqrt{(10\mu)(5p)}$$

$$\therefore \quad R_R < 1.9 \text{ k}\Omega$$

$$R_R \approx \frac{L_{CH}}{W_H K_P'(v_O - |V_{TP0}|)}$$

$$= \frac{190n}{W_R(40\mu)(1.8 - |-700m|)} < 1.9 \text{ k}\Omega$$

$$\therefore \quad W_R > 2.3 \text{ }\mu\text{m} \quad \rightarrow \quad W_R \equiv 3 \text{ }\mu\text{m}$$

6.5. Switched Diodes

Asynchronous power supplies use diodes to drain L_X. Since diodes block reverse current, L_X transitions into DCM automatically (without a ZCD) when i_O falls. The drawback is the power they burn with the 500–800 mV they drop when they conduct $i_{L(D)}$.

Switched diodes are transistors that behave like ideal diodes. They close when their voltage is positive and open when their voltage reverses. This way, with millivolts across them when they conduct $i_{L(D)}$, they dissipate little power. And with current flowing in one direction only, $i_{L(D)}$ cannot reverse and burn unnecessary power.

A. Low Side

Buck-based supplies use a *low-side diode* D_L to drain L_X. In Fig. 35, the low-side comparator CP_{DL} switches M_L like an ideal D_L. CP_{DL} trips v_G high or low to close or open M_L when v_{SW} falls under or climbs over v_L.

In practice, however, CP_{DL}, the inverter, and gate driver require time t_P to react. M_L's body diode should conduct i_L across this t_P. This is why M_L's body connects to v_L, so i_L can flow through the body diode.

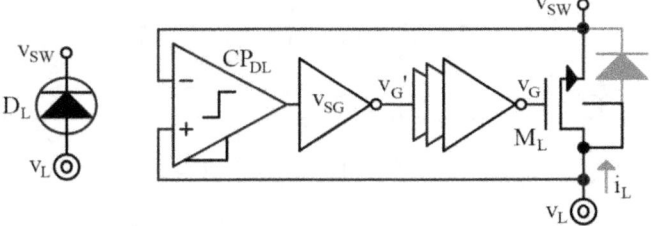

Fig. 35. Switched low-side diode.

B. High Side

Boost-based supplies use a *high-side diode* D_H to drain L_X. In Fig. 36, the high-side comparator CP_{DH} switches M_H like an ideal D_H. CP_{DH} trips v_G low or high to close or open M_H when v_{SW} climbs over or under v_H.

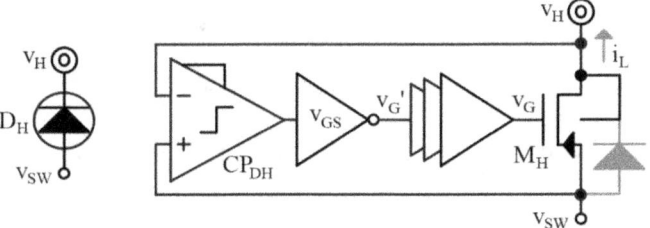

Fig. 36. Switched high-side diode.

In practice, however, CP_{DH}, the inverter, and gate driver require time t_P to react. M_H's body diode should conduct i_L across this t_P. This is why M_H's body connects to v_H, so i_L can flow through the body diode.

6.6. Starter

A. Shutdown

A *shutdown* command should open all power switches. This way, the drain diodes deplete L_X into C_O and the *load* R_{LD} and R_{LD} discharges C_O. So i_L and v_O end at zero.

In buck–boosts, *ground* and *output diodes* D_{DG} and D_{DO} in Fig. 37 *drain* L_X into C_O and R_{LD}. And as R_{LD} discharges C_O, R_{LD} and R_L drain leftover energy that L_X and C_{SW}'s exchange. So v_{SW}'s oscillations shrink with v_O as long as D_{DO} conducts and cease when R_L burns E_{LC}.

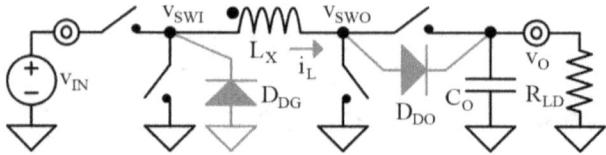

Fig. 37. Shutdown buck–boost.

The buck excludes the output switches that connect L_X to v_O. So the shutdown process is the same, but without D_{DO}. D_{DG} in Fig. 38 drains L_X into C_O and R_{LD}, R_{LD} discharges C_O, R_{LD} and R_L drain leftover energy that L_X and C_{SWI} exchange, and v_{SWI} oscillates until R_L burns E_{LC}.

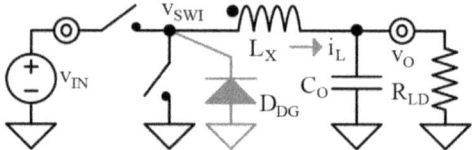

Fig. 38. Shutdown buck.

The boost excludes the input switches that connect v_{IN} to L_X. So the shutdown process is similar, but without D_{DG}. D_{DO} in Fig. 39 drains L_X into C_O and R_{LD}, R_{LD} discharges C_O, and R_{LD} and R_L drain leftover energy that L_X and C_{SWO} exchange.

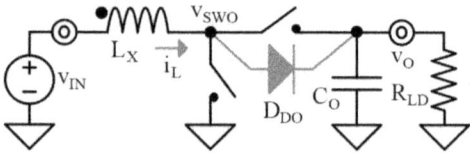

Fig. 39. Shutdown boost.

But since L_X eventually shorts, v_{SWO} oscillates until D_{DO} clamps v_O to a diode below the v_{SWO} that v_{IN} sets. So after draining L_X, L_X shorts and i_L climbs up to an ohmic translation of v_{IN} and v_{DO}, which $R_{L(DC)}$ and R_{LD} set. In short, shutdown sets i_L and v_O to

$$i_{L(SHUT)} = \frac{V_{IN} - V_{DO}}{R_{L(DC)} + R_{LD}} \tag{55}$$

$$V_{O(SHUT)} = V_{IN} - i_{L(SHUT)} R_{L(DC)} - V_{DO}. \tag{56}$$

To shut v_O and i_L down to zero, D_{DO} would need to open somehow.

Example 17: Determine $v_{O(SHUT)}$ and $i_{L(SHUT)}$ for a boost when v_{IN} is 2 V, R_L is 250 mΩ, R_{LD} is 4 Ω, and I_S is 50 fA.

Solution:

$$i_{L(SHUT)} = \frac{v_{IN} - V_t \ln(i_{L(SHUT)}/I_S)}{R_L + R_{LD}}$$

$$\approx \frac{2 - (26m)\ln(i_{L(SHUT)}/50f)}{250m + 4} \qquad \therefore \quad i_{L(SHUT)} = 290 \text{ mA}$$

$$v_{O(SHUT)} = v_{IN} - i_{L(SHUT)} R_L - V_t \ln\left(\frac{i_{L(SHUT)}}{I_S}\right)$$

$$= 2 - (270m)(250m) - (26m)\ln\left(\frac{270m}{50f}\right) = 1.2 \text{ V}$$

B. Startup

Starters wake power supplies from their shutdown state. They first wake the comparators and amplifiers that monitor and manage the system. And once armed, these blocks control and command i_L to supply the load.

Since C_O is usually very high and $v_{O(SHUT)}$ is well below the targeted v_O, the feedback action of the system is to supply a very high i_L. Unfortunately, such a high inrush of current can burn components. So once blocks are ready, the starter should impede their action until i_O and v_O near their targets.

Over-current protection (OCP) is one way to keep i_L from damaging the system. OCP disables the power stage when i_L reaches $i_{L(MAX)}$ and re-enables it when i_L falls below $i_{L(OK)}$. This on-and-off process repeats until i_O and v_O reach their targets. Note OCP duals as *short-circuit protection*.

Limiting the *duty-cycle command* d_E' that energizes L_X during *startup* is another way. But since d_E' also limits v_{IN}'s translation to v_O, $d_{E(START)}$ should be greater than the d_E needed to set v_O, but lower than 90% or 95%. Or $d_{E(START)}$ can ramp slowly to d_E. This way, i_L never reaches $i_{L(MAX)}$.

The reference voltage can also *slow-* or *soft-start* the system. v_R can start at zero and ramp slowly to the v_R needed to set i_O or v_O. This way, the feedback controller ramps i_L with v_R slowly.

7. Summary

Switched-inductor power supplies are mixed-signal systems. They embed analog, analog–digital, and digital functions that set and regulate i_O or v_O. Some of the building blocks needed for this functionality are sensors, amplifiers, comparators, timers, digital logic, and switch controllers.

Current sensors normally hinge on resistance. Using switch and inductor resistance already in the network for this purpose is more power efficient than adding resistance. This resistance, however, varies widely with temperature and fabrication runs. Sense transistors often offer more favorable tradeoffs in this respect. Their weakness is mismatch.

Voltage dividers sense and translate voltages well because resistors usually match well. Paralleling a capacitor across the top resistor is an easy way of inserting a phase-saving zero–pole pair. Combining the voltage divider with the stabilizing error amplifier is also possible.

Amplifiers and comparators compare analog inputs and output a digital output that indicates which is higher. Comparators are basically amplifiers without stabilizing features. Adding flip flops or positive feedback establishes hysteresis. And paralleling transconductors adds inputs so the output trips with the polarity of their sum.

Constant-time loops rely on timing blocks to operate. Capacitors are essential here. In sawtooth generators, current into a capacitor ramps a voltage that a flip flop resets. Sawtooth and one-shot oscillators ramp a voltage that a comparator resets. And a flip flop in the one shot can interrupt the oscillations to pulse once or any number of times.

Digital blocks control switching events. SR flip flops use digital gates and positive feedback to decouple on–off commands. Inverter chains with increasingly larger stages drive large power switches. And dead-time logic keeps adjacent power switches from shorting their inputs.

Switch blocks help manage switching events. Supply-sensing comparators are useful in this respect. They help zero-current detectors invoke DCM operation and switched diodes behave like ideal diodes. They can also trigger the ring suppressor, which subdues DCM oscillations.

The starter is also important. It keeps initial shutdown conditions from spiking i_L. Current and duty-cycle limiters are useful guardrails during startup. Ramping the reference that sets the output is another safety measure that can soften the impact of power-up on i_L.

www.ingramcontent.com/pod-product-compliance
Lightning Source LLC
Chambersburg PA
CBHW070827220526
45466CB00002B/770